D1217628

Illustrated Guide To Basic Electronics:

With Useful Projects And Experiments

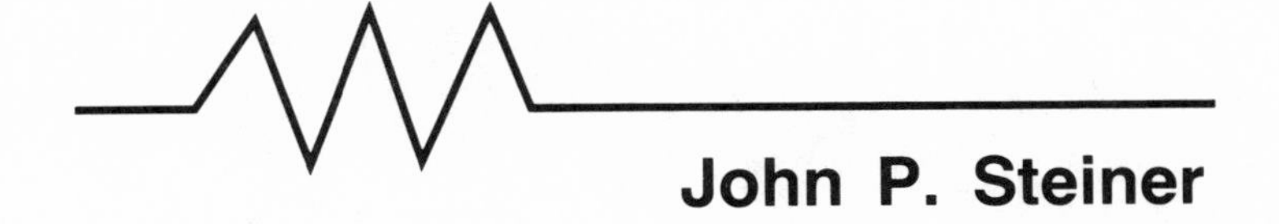

John P. Steiner

Prentice-Hall, Inc.
Business and Professional Division

Englewood Cliffs
New Jersey

Prentice-Hall International, Inc., *London*
Prentice-Hall of Australia, Pty. Ltd., *Sydney*
Prentice-Hall Canada, Inc., *Toronto*
Prentice-Hall of India Private Ltd., *New Delhi*
Prentice-Hall of Japan, Inc., *Tokyo*
Prentice-Hall of Southeast Asia Pte. Ltd., *Singapore*
Whitehall Books, Ltd., *Wellington, New Zealand*
Editora Prentice-Hall do Brasil Ltda., *Rio de Janeiro*

Library of Congress Cataloging in Publication Data

Steiner, John P.
Illustrated guide to basic electronics.

Includes index.
1. Electronics I. Title.
TK7816.S73 1984 621.381 83-3422

ISBN 0-13-450510-7
ISBN 0-13-450502-6 {PBK}

Editor: George E. Parker

Printed in the United States of America

A Word from the Author on the Unique, Practical Value This Book Offers

The wide range of practical material in this book is aimed directly toward the individual who is fascinated by the electronic miracles that change our lives almost on a daily basis. As an experimenter or hobbyist, you would like to learn more about electronics (particularly from a practical standpoint) but have no intention of going back to school for this purpose. In the following pages, you'll find everything you need in order to gain a functional understanding, while rapidly increasing your ability and skills in electronics.

We have included many illustrations and photos keyed to the text that will help you to identify equipment, tools, and components. By hands-on experiences, you will not only learn about theoretical concepts of electronics, you will actually be able to apply what you have learned to the real world of electronic circuits. This book also provides the information needed to help you select the right tool for the job at hand. It has been said many times before that purchasing an incorrect tool, because of price, is false economy. The information on tools, alone, will save you many times the cost of this book in avoiding frustration, as well as saving money.

One of the most versatile, yet inexpensive, pieces of test equipment in the electronics experimenter's shop is the multimeter. If you do not already have one, you will need a meter that is sufficient for your needs, yet does not have more capability than you require. There are a lot of meters on the market, with a wide range of functions and price. The multimeter, in its basic form is able to measure the amounts of voltage and current in a working circuit. It will also measure resistance, or the ability of a circuit or component to oppose current flow. Helpful hints are provided that will help you make sure you get the most equipment for your money, and when it is time to upgrade, you will be able to choose among alternatives.

Also included are complete operating instructions for the multimeter, as well as some hints on how to check out used meters before you buy. These are simple tests that can be carried out in a few minutes and will help to assure you that the used meter you buy will work properly when you get it home. You will learn to use your multimeter to test diodes, transistors, capacitors, resistors, and other components. The mysteries of these components will be explained in a manner that makes the concepts easy to understand.

One of the most revolutionary developments in modern history occurred in 1949, when three Bell Telephone Laboratories scientists invented the transistor. This new technology, dubbed solid state because amplification was occurring in a solid piece of material, has changed everybody's lives. You will come to understand the transistor and be able to explain its functions.

The inception of the integrated circuit (IC) was no less revolutionary than the invention of the transistor. Almost every electronic device, from your car radio to your microcomputer, uses ICs by the handful. Indeed, if you own a 1980 or newer automobile, the chances are that major engine functions are provided by an integrated circuit whose job is to retune your engine for maximum performance and economy at least 10,000 times each minute, all this while testing each function and storing information about defective engine components. The miracle children of the space program are explained in a practical manner, and they will have many applications around your home and shop.

Even if you know about these components, you still cannot put them to practical use unless you interconnect them properly. You will not only learn how to interconnect components by the time-honored technique of soldering, you will also learn modern wiring techniques that have replaced soldering. The techniques are explained in simple step-by-step fashion, and you will be able to decide which technique is best for your particular application.

Have you ever been discouraged by long chapters on electronics math? If so, you will find the chapter in this book refreshing. There is no more math than is absolutely necessary for understanding practical electronic circuits, and a simple pocket calculator will provide all the assistance you need to completely understand and apply appropriate math concepts to electronic circuits.

Chapter 10 contains twelve electronic projects that not only are interesting to build, but will become useful equipment around your electronics shop, and your home. Each project helps to reinforce certain important concepts in electronics. A few are just fun to build... and fun to use when completed. The final project completes a working signal tracer, a useful piece of test equipment that will become an asset to your workbench. The tracer provides, at minimal cost, many of the troubleshooting techniques one usually associates with the oscilloscope, which is a useful but expensive testing device. The signal tracer provides a means of entering a working circuit, and listens for the signals that should be present in each stage of a radio or amplifier. Complete instructions are included for using the tracer properly, and, even if you buy all new parts, it should cost under $25.00—considerably less if you have a large junk box of components, or access to used parts.

All projects use readily available parts, and include a cross-reference to several sources so that you may find local suppliers, and eliminate mailing delays. These projects have all been built and tested by other beginners under my direction, and can be depended upon to work reliably, as long as the step-by-step instructions are followed. Each project also includes a troubleshooting chart to assist you, in case the project does not work the first time you turn it on.

As a hobbyist, you have probably tried to repair some form of electronic equipment, whether it was a simple transistor radio or a complete cassette stereo tape deck. Did you have to give up on it, or could you calmly apply your electronic knowledge to locating the problem? If you apply the techniques learned in the troubleshooting chapter, combined with the useful collection of flow charts included, you should be able to troubleshoot just about any piece of electronic equipment or project.

This book's usefulness does not end when you have finished reading it. There is a large index in the back that will help to ensure that you will use this book as a daily reference tool for many years to come.

In the world of electronics, one never ends the learning process, so take your time and enjoy yourself. When we succeed in increasing your awareness of the miracle of electronics, and thereby developing greater enjoyment of this hobby, our objective will have been achieved.

John P. Steiner

Acknowledgments

I am grateful to the many individuals who assisted me in producing this manuscript. First, Leslie Pierce spent many long hours preparing the artwork for the illustrations. Ken Christiansen read the manuscript as it was being developed to locate any technical errors; the manuscript is far better because of his input. Bob Gilbertson assisted in the project design and selection. James Heising printed many of the photos included. Richard Stenberg and Dennis Stormoe also assisted in developing the photos and illustrations. Most of all, my wife, Lynn, and my family have been of great assistance by allowing me to spend many weekends working on the manuscript. Without their cooperation, I could never have finished the task.

Contents

Electricity: What It Is And How To Measure It

1

Have you ever looked at a water tower and thought of all the water that was stored there? Or looked over a dam and thought about all of the potential for damage should the water be somehow let loose? That potential to release energy in small amounts or large is similar to the potential of voltage—the pressure, sometimes called electromotive force, is the potential to do work, and provides the push to allow the flow of electrons that actually do the work. The name voltage was given to this unit to honor an early electrical pioneer by the name of Volta.

How Electrical Pressure and Current Are Related

It is easy to be overwhelmed by the concepts of voltage and current when learning the basics. Unfortunately, people have used these terms interchangeably in conversation, when in truth they are two totally different but related phenomena. While voltage is potential, or pressure in a circuit, current is the actual movement of charged particles within conductors in the circuit. These particles are called electrons. For now, just remember that current flow is the movement of electrons in an electrical circuit.

To understand the concept of voltage more thoroughly, let's return to the comparison of water pressure to electrical pressure. The water faucets and electrical outlets in your home are, in many ways, similar. Each water faucet has the potential to release energy, and each outlet has the same potential. By turning on a faucet or flipping a switch, you release the potential energy in a controlled manner. If the energy should get out of control, the potential for damage is readily apparent. The voltages that are standard in most U.S. households are 220 volts and 110 volts alternating current. We will explore the concepts of alternating and direct currents later. The range of voltages in common use is quite large. Watches and calculators often are designed to operate on pressures as low as one volt, while certain components in television sets require 25,000 volts or more to operate. Electrical storms often generate potentials of several millions volts.

You might ask how this potential is created. There are six major sources of electricity. A brief explanation of each source is included for future reference, though a complete discussion of electrical generation is beyond the scope of this book. The six sources of electricity are chemical reaction, heat, light, magnetism, friction, and pressure.

The battery uses chemical reaction of two dissimilar materials to cause electrons to flow. There are many types of batteries, which are classed by their materials and construction. Some batteries are meant to be used once and discarded. These are primary batteries. The primary, or nonrechargeable, dry cell is commonly found in flashlights, transistor radios, and many other types of portable electronic equipment. These dry cells are made of carbon and zinc, or alkaline and magnesium. They contain a relatively dry paste chemical electrolyte that activates the generation of potential between the cell's output terminals.

These cells are sealed, and if you try to charge them, the gases released by the charging process might cause the internal pressure to puncture the case. The cell will rupture or explode, causing possible injury—and a big mess to clean up.

As you might have noticed, I have not called these devices batteries. Though the term is commonly used to designate a cell, the term "battery" is actually meant to refer to more than one cell. In other words, a flashlight typically needs a battery of two dry cells.

The battery in your car is formed of six wet cells. The wet cell is so called because it uses a liquid electrolyte, an acid that reacts to the lead plates to generate a potential. The wet cell is a secondary cell that can be recharged again and again. Other types of secondary cells are also available. The most common is the ni-cad battery. This cell uses nickel and cadmium to provide the difference of potential. These cells are commonly used in portable receivers, walkie-talkies and other high current-demand equipment, when reliable, reusable power sources are required.

Heating a special combination of materials will cause a difference of potential to be generated. These conductive materials, when heated, develop a potential at their respective outputs. The devices that exhibit this principle are called thermocouples. They should not be confused with thermostats, which are on-off switches that open and close depending upon the temperature. An application for a thermocouple might be as a sensing device for an electronic thermometer.

One of the long-term solutions to our energy shortage is electricity generated from light. The photocell shows promise that, on a sunny day, most of the needed electricity will be provided by the sun. The process of photoelectricity is based upon light reacting inside a crystalline material. When light energy strikes these materials, the light will dislodge electrons from their natural atomic orbits, and a voltage or pressure will be developed.

Probably the most common method of generating electricity is by magnetism. The electric generator uses the effect of magnetic fields to induce a potential in a coil of wire—a potential due to the lines of magnetic force, called flux, intercepting the conductors in a coil. Magnetism is a powerful force, as anyone knows who has seen an electromagnet pick up a junk automobile. The magnetic field actually causes electrons in a conductor to travel through the conductor. Much more will be said on the topic of magnetism in Chapters 2 and 6. The biggest generators are used in commercial generating power plants that provide nearly everyone with their electricity. Massive generators may be powered by diesel or gasoline engines, or by steam pressure using huge boilers. Another magnetic generator is powered by the hydroelectric power plant. These plants, built on rivers, use the flow of water to turn the generator, thus producing electricity.

Nearly everyone has walked across a carpet and touched a door knob, only to receive a surprising shock. This form of electricity, called static electricity, is generated by friction, in this case, the friction of shoes rubbing against carpet. The friction of moving air masses in thunderstorms also generates electricity. The release of this electrical energy is sudden, and immense heat, light, and noise are generated in the lightning bolt.

Certain materials, when stressed, will develop a potential. An example of this is found in the ordinary record player. In many phonographs, the needle is attached to one of these materials, usually ceramic. The vibration of the needle in the record groove flexes the ceramic, which develops a potential in exact relationship to the music recorded in the grooves. The small potential is amplified many hundreds of times and fed to a speaker. This is an example of electricity generated by pressure.

All of these sources provide the potential for electrical energy; however, having potential and using it are two different things. Just as the water rests behind a dam, or in a water tower, voltage is there to deliver energy as the demand arises. When you turn on the faucet, water can flow at a rate of several gallons per hour, or more. When you turn on a switch, current flows. The rate of current flow is also expressed as units per time. While the flow of gallons of water is actually thousands of drops of water, electricity flowing consists of millions upon millions of electrons. Since so many electrons are flowing, the number is expressed in a quantity much easier to grasp.

This quantity is called the coulomb, and the number of

electrons in a coulomb is 6,240,000,000,000,000,000. For those who are familiar with scientific notation, this is expressed as 6.24×10^{18}. Obviously, it is much easier for us to talk about coulombs when expressing quantity. Though an improvement, the coulomb becomes difficult to work with, as after a short period of time, the number of coulombs becomes unmanageable. As a result, it is much easier to talk about a rate of flow. Water flow is often measured in gallons per hour, and current flow is measured in coulombs per second. This convention keeps the numbers manageable. The coulomb per second has been given a simpler name, the ampere, also named after an early electricity pioneer. Simply stated, current flow in an electric circuit is expressed in amperes, or amps for short.

The rate of current flow is directly proportional to the amount of pressure of voltage available to an electric circuit. When there is greater pressure in a circuit, more electrons will move past a given point in a circuit. The faster the electrons move, the more current that flows in a circuit. Most circuits draw relatively little current. Transistor radios seldom draw more than 0.1 ampere or 100 milliamperes. A milliampere is 1/1000 of an ampere. Automobile batteries deliver considerable current when starting a car. Many large automobiles need over 200 amperes of current flow to turn the starter motor. This brings up a concept that may have already occurred to you. If a transistor radio operates on twelve volts, and an automobile starter motor also operates on twelve volts, what keeps the transistor radio from needing 200 amps? The answer to this question lies in the nature of the radio and the motor themselves. Each device contains conductors and insulators. Each contributes a single factor that is related to a current flow. This factor actually opposes the flow of current in a circuit, and is called resistance.

Opposition to Current Flow

All materials present a certain opposition to the flow of electrons. Electron flow requires the dislodgment of electrons in the atomic structure of a material. How easily a material allows electrons to be dislodged and moved, one atom to the next, is the greatest factor in the conductivity of a material. If the atomic structure allows current to flow readily, it will offer little opposition to current flow. These are the materials found in the starter motor. The transistor radio, on the other hand, contains materials in its circuits that are much more resistant to the pressure of voltage that is applied. As you can see, when opposition to current flow is great, the amount of current flow is small. The action of resistance is actually opposite to the action of voltage in a circuit. In other words, current and resistance are inversely proportional. Just as electrical pressure and current have names, resistance has been given the name of an early pioneer, and is expressed in ohms. Resistance varies greatly depending upon the type of material involved, and commonly found resistances in electronic circuits vary from less than one ohm to well over a million ohms. Values of resistance above a thousand ohms are often expressed in kilohms, and one kilohm is equal to 1000 ohms. Resistances above one million ohms are usually expressed in megohms, and a megohm is, as you have probably already guessed, equal to one million ohms. The relationship between voltage, current, and resistance is the foundation of electricity and electronics, and is expressed in the mathematical concept of Ohm's law. This law is the most important concept that an experimenter will have to use in building electronic circuits; it will continually assist the experimenter in deciding which value of resistor or other component to use when building or designing a circuit. Probably the best thing about Ohm's law is that the relationships are so easy to understand. Basically, the law states that one volt will cause one amp of current to flow through one ohm of resistance. If the voltage is doubled, the current will automatically double, (remember voltage and current are directly proportional). By doubling the resistance, you can cut the current in half, (again remember, current and resistance are inversely proportional). The experimenter will use these laws and concepts many times over, while deciding which components to use. Ohm's law, and its application to your experiments, is covered in Chapter 4. We will refer to this concept again, and expand upon it when it is needed.

It is the experimenter's responsibility to limit the amount of current that can flow in a circuit to safe levels. You must meet this responsibility by judicious use of the concepts you will be shown. By controlling the voltage and resistance in a circuit, you will be meeting the above requirements. The experimenter must connect components in his circuits so that the current is directed where he wants it, and directed away from places where it is not wanted. Two items in the experimenter's arsenal to accomplish this end are the conductor and the insulator. You may have already thought about why an insulator will not allow electricity to flow through it, yet a conductor will allow current to pass through quite readily. The difference is in the atomic structure, as was stated earlier. Let's take a look at how the structures differ.

How Conductors and Insulators Differ

If you could look into a conductor or insulator with a powerful microscope to see the actual structure of the atoms, you would see a structure that contains a relatively dense center core, called the nucleus. This nucleus contains positively charged particles known as protons and particles that contain no charge called neutrons. The basic charge in a normal atom's nucleus, or central core, is positive. Orbiting around this central core are electrons, negatively charged particles that are found in several orbital locations. The number of electrons in most materials equals the number of protons, thus the net charge of the atom in total is neutral. All materials contain atoms of differing internal structure basically affecting the number of protons and electrons. There is a definite pattern to the orbiting electrons that varies with the material.

Though this pattern is important in identifying the material, its exact structure is beyond the scope of this book, and is not really necessary to understand electricity. The orbiting electrons that are farthest from the nucleus, however, are quite important to current flow. An atom that has relatively few electrons in its outer orbit, compared to its capacity to hold electrons, will give up those loose electrons quite easily. A small amount of electrical pressure will literally push loose electrons from their orbits. The free electron may collide with another electron, knocking it out of orbit. The free electron then is captured by the new atom, and the electron that was knocked loose becomes another free electron. Free electrons are attracted to the positive charge, and move toward the positive electrical terminal. This process rarely knocks more than one or two electrons from an individual atom's orbit, and they are soon replaced by electrons supplied from the battery. The pressure or voltage applied to the conductor determines the amount of interaction, and the amount of current flow in a given material.

The exact type of material also determines the amount of electrons to be freed. The materials whose atoms give up their loose electrons most readily include silver, copper, gold, and aluminum. Gold and silver are far too expensive to be used as electrical conductors in wire, so copper is the metal most used. Aluminum is nearly as good a conductor as copper, and is less expensive, though aluminum is rarely used in electronic circuits due to the fact that soldering and interconnecting aluminum wires is difficult. These metals have one thing in common—they give up their electrons very easily. When high conductivity is required, silver is used in small quantities. Gold is used when corrosion resistance is required. The elements are usually bonded to switch and relay contacts, and used inside integrated circuits. Techniques have been developed by industry to reclaim the small bits of gold from worn-out parts to be used again on new components.

The atomic structure of an insulator is similar to that of a conductor; however, unlike conductors' orbits, the outer orbits of insulators are full or nearly full of electrons. These electrons are bound quite tightly to the atom and much pressure must be applied to knock electrons loose from their atoms. What this means to you as an experimenter is that all materials will conduct if you apply enough pressure or voltage. In other words, you must be cautious around high voltage supplies, as insulators will indeed conduct, given the right circumstances. Most wire and cable that can be purchased today has a voltage rating of at least 600 volts, and any project that operates using less than this will not demand special consideration of the insulation being used. Television sets whose picture tubes need 25 to 35 thousand volts to operate must have insulators on their conductors that will not conduct at potentials of up to 50 kilovolts. Using cable with inadequate insulating properties may lead to shock or fire as the pressure of the high voltage turns the insulator into a conductor. Typical insulating materials are rubber, glass, plastic, and ceramic.

There is one other class of materials that should be mentioned here. These materials contain fewer electrons in their outer shells than insulators but more than conductors. Materials that have this characteristic are called semiconductors. The semiconductors that are most commonly found in electronics are silicon and germanium. These materials are the wonder elements of electronics, because without them there would be no transistors, diodes, or integrated circuits. By selecting the proper materials, it is possible to obtain our objective of allowing current to flow only where we want it, while restricting it from circuits where we do not.

One type of material that has more solidly bound electrons than most conductors is carbon. This is the element that is used in resistors. The manufacturers of resistors carefully control the amount of carbon and impurities deposited in the resistor. By controlling the amount of carbon and impurities, any value of resistance can be obtained. There is more detail on this process in Chapter 2.

Pressure applied to a conductor will cause electrons to flow in a circuit quite easily. The actual work being done is done by the movement of electrons. In many circuits, such as in lights and heaters, the direction of current flow is immaterial. As a result, current is available in either of two forms, alternating current or direct current. Let's look at the advantages and disadvantages of the two forms.

Differences Between AC and DC

Look at the circuit in Figure 1–1. This is a direct current or DC circuit. When you close the switch, current will flow because of the potential energy stored in the battery. Electrons will flow out of the negative battery terminal and into the switch. After leaving the switch, the electrons will flow through the bulb and, from there, into the positive battery terminal. The resistance of the bulb limits the current flow out of the battery to an acceptable level, and, at the same time, this resistance makes it difficult for electrons to move freely. The resulting collisions of electrons into the more tightly bound atoms of bulb filament generate heat so intense that the bulb begins to glow. If we were to reverse the battery terminals, putting the negative terminal on the bulb, and positive terminal on the switch, the results would be the same. Keep this in mind as we explore the nature of alternating current. When Thomas Edison invented the light bulb, many people regarded it as one of the greatest inventions in modern history. While the light bulb is indeed a marvelous invention, it would be useless without the power generating station that Mr. Edison invented to supply power to his light. The electric power generating

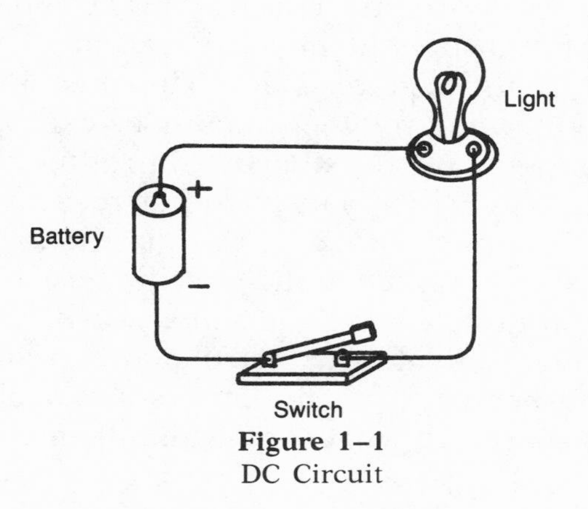

Figure 1–1
DC Circuit

station that supplied voltage to these early light bulbs was a direct current generator. It worked satisfactorily as long as there were few circuits to connect, and they were relatively close to the generating station. Problems became apparent when it was realized that the added load of each new customer caused the increasing need for larger and larger conductors at the output of the generating station. Also greater current meant larger voltage drops across the long length of wire, which could lead to brownouts.

Another early pioneer of industrial technology, probably remembered more for his work on the locomotive, George Westinghouse, determined that a solution to this problem could be found in alternating current generators, or alternators. An alternator, instead of generating a single polarity voltage, actually generates a voltage that continuously varies in potential as well as polarity. Figure 1–2 shows the action that occurs in an AC circuit. In A, the arrow shows the direction of current flow. An instant later, however, current flow changes direction and flows the other way, as shown in B. This action occurs many times each second. In the United States, the change in polarity occurs 60 times each second, while in Europe, the standard is 50 times a second. Other values could have been used, indeed AC circuits have been

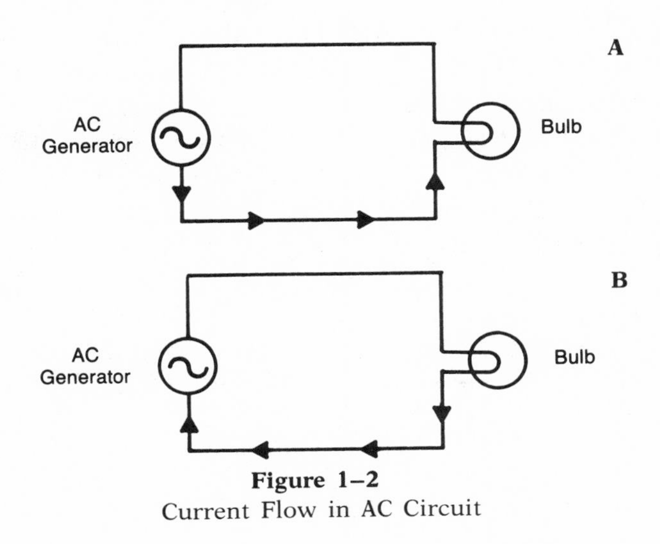

Figure 1–2
Current Flow in AC Circuit

built that change polarity well over one billion times per second.

As has already been said, the motion of electrons is there, so the bulb will light, however this seems a complicated procedure, when a DC generator will do the same thing. The practicality will become apparent when the nature of transformer action is understood. Transformers are electric components that will be discussed in greater detail later; however, for our present purposes, a transformer has the ability to change voltage and current levels. Refer to Figure 1–3. A transformer can be designed to work at any voltage and current level desired. The power company can produce AC at 11000 volts with a current of one amp, and send it down the power lines. This little current requires very small wires, and thus costs less. At the home, a transformer converts the power line voltage and current to 110 volts at 100 amps. While this is a simplified explanation, the section on transformers in the next chapter will explain the process in greater detail.

The process of voltage conversion is dependent upon AC because transformers will not work in DC circuits without some method of changing DC to AC. It would seem that if AC is this versatile, that DC would have long ago been discarded, but there are many components that require DC to operate. One of the most needed circuits in electronic equipment is the power supply. This supply must first transform the AC to the desired voltage value, then change it to DC so that the project will work properly. Chapter 7 discusses this in enough detail so that you will be able to design your own supply.

Whether you are working with AC or DC, you must be

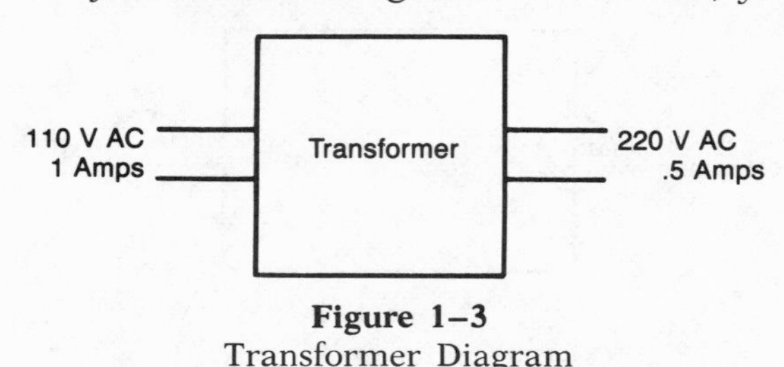

Figure 1–3
Transformer Diagram

aware of the voltage and current levels in circuits you design or build; you must have an accurate way to measure voltage, current, and resistance. If you already own a meter, you already know of its usefulness. If you don't have one, the next section will help you to select a reliable and accurate meter, and to continue to use it properly and safely.

Choosing and Using a Multimeter

In the earlier days of electricity, persons interested in measuring voltage, current, and resistance had to purchase or build three different meters. As an experimenter, you will need to be able to measure each unit, however, most meters today measure all three values. This is done by inserting probes into different sockets, by changing a switch position, or both. Single-purpose meters are still available for dedicated use. For example, you can buy a 0-to-150 volt AC meter that you connect to your wall outlet to keep track of the variations in your line voltage. Equipment such as transmitters requires the monitoring of internal current at regular intervals, therefore the manufacturer includes an appropriate ammeter that will continuously monitor this important specification. The single-purpose meter, however is usually inappropriate for the experimenter. Selecting an appropriate meter can be confusing to the novice, unless he or she has obtained a few basic facts beforehand.

Multimeters are available in analog and digital types. Analog meters contain metered scales, while digital meters have a readout similar to a digital clock. Both types will be discussed separately, starting with the analog style. Another important characteristic, especially for the novice, is the amount of internal meter protection included. The well-protected meter shuns accidental overloads that would send a lesser meter up in smoke.

The three most important parameters, or characteristics, of multimeters are: percentage of accuracy, ohms-per-volt rating, and number of ranges. Each of these parameters will be discussed separately, and when you choose a meter, you will have to find an agreeable blend among them. You must also keep your eye on another important parameter, cost.

The characteristic that is easiest to identify, and interpret, is percentage of accuracy. Readily available commercial multimeters have a percentage of accuracy that ranges from less than 1% to 10% of full scale readings. With analog meters, though, this percentage can be deceiving. What the specification doesn't always state is that this percentage is valid only at full scale. This means that as the meter pointer falls toward the left end of the scale, the accuracy of the meter decreases. In the lower half of the scale, the reading might be only 20% accurate for a meter rated at 10%. For example, a full-scale accuracy of 1 percent means that on the 100-volt range the meter will always read between 99 and 101 volts when a potential of 100 volts is applied. It will also read within 1 volt of the true value at any other point on the scale. When reading meters, then, it is a good idea to select a range that causes the meter to swing as far to the right as possible. More on this later.

The ohms-per-volt rating is closely related to accuracy. As you already know, a large resistance allows less current flow through it than a small resistance. When meters are measuring voltage, they are actually introducing an extra path for current to flow. When large currents flow through a voltmeter, the circuit being measured is upset. The voltage changes because you have inserted a meter, which might cause a voltage that seems to be out of specification. As a result, you may waste time looking for a nonexistent circuit problem.

Though all meters will slightly affect the circuit under test, meters with a higher ohms-per-volt rating upset the circuit least. For now it is not important to completely understand the concept of ohms-per-volt. Choosing a meter with a rating of 20,000 ohms-per-volt minimum is a good rule of thumb. The more expensive meters generally have a higher rating in this important specification. Meters that have higher than 50,000 ohms-per-volt rating are amplified. In other words, they contain electronic circuits that improve this characteristic. These circuits add to the cost and complexity of the meter, but are usually worth the extra money for an avid experimenter. Better quality meters can have a rating as high as one megohm-per-volt or more.

The number of ranges is important in selecting a meter. As a rule, those meters with more ranges will have more opportunity to ensure that the pointer will be in the upper half of the scale; this will give the most accurate reading. Ranges are selected in two different ways. The easiest to use, and most complex internally, is the range switch. If you use a single switch, all voltage, current, and resistance ranges can be selected in a single step. In order to save on internal wiring, and its resulting extra cost, some meters will have separate sockets, in addition to the range switch, that the meter probes are plugged into to select less frequently used ranges. Figure 1–4 is a photograph of one meter of this type.

Photo courtesy of Radio Shack, a division of Tandy Corp.

Figure 1–4
A Typical Analog Multimeter

The meter probes are normally inserted into the jacks marked COM (common) and V-Ω-A (volts, ohms, and amps). If you expect to use the 10A (10 amp), or the 1000V (1000 volt) scale, you must select the appropriate jack, as well as the correct meter switch position. Trying to measure more than 250 volts, for example, without selecting the correct jack for the probe will probably result in a shock hazard, as well as damage to the meter. Some less expensive meters might do away with the range switch altogether, requiring you to select the proper range by inserting the probes into the correct sockets. Try to avoid this type of meter. It is annoying to be continually changing the probe leads around, and forgetting only once can destroy a meter. Usually meters of this quality have an ohms-per-volt rating that is unacceptable for measuring voltages in solid-state circuits.

Unfortunately, the easiest and best meters for the experimenter to use also happen to be the most expensive. Electronics is not the only area in which this is true; photography is another example of an area where the most expensive equipment produces the best results most simply. One parameter that is not to be overlooked is meter protection. Meter movements are fragile and sensitive devices. A good meter movement will deflect to full scale, all the way to the right, with only 50 microamperes of current flowing through it. Currents greater than this will damage or destroy the meter movement. Meters usually have some form of protection, however, inexpensive meters are the least well protected. At the very least, the current ranges should be fuse protected. The better-protected meters contain fast-acting relay circuits, controlled by electronic circuitry, that sense an overload, and instantly disconnect the meter movement from the test leads. These types of meters must be reset after an overload occurs. The money saved by purchasing a meter lacking in protection will be lost the first time you make a mistake. If the choice arises, where you must decide among more ranges, and a better protection circuit, take the better-protected meter.

The newest type of meter available is the digital multimeter (Figure 1–5). In a few short years, the cost of these

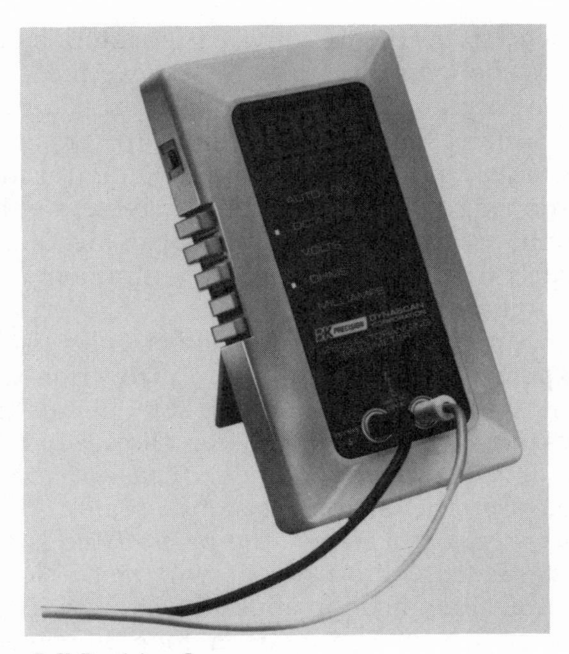

Photo courtesy B-K Precision Group

Figure 1–5

An Auto Ranging Digital Multimeter

easy-to-read meters has dropped from several hundred to well under a hundred dollars. Their advantages are in ease of use and accuracy. As you will soon see, deciphering a complex meter scale is sometimes difficult for the beginner. A digital reading is read directly, for example, if the range switch says 50 volts, and the dial reads 25.5 volts, this is easily interpreted. There is no multipling or decimal point adjustment. This direct reading also eliminates the scale accuracy problem of the analog meter. Digital meters typically have a percentage of accuracy of better than 3%, and this accuracy is constant on each scale, at all points, so that lower ranges are necessary on a digital meter. The resistance rating on a digital meter is usually a minimum of 10 megohms, allowing for highly accurate readings.

Another advantage of the digital meter is that most of them are designed with an auto-polarity function. This means that negative voltages will be indicated with a minus sign to the left of the reading. An analog meter, when connected to the wrong polarity, will move the meter to the left of zero. To read the negative voltage, you must either reverse the meter leads or, on some meters, change the position of a lead polarity switch. Some digital meters also contain an auto-range function. These meters will have no range switch, only a volt-ohm-amp select switch. The automatic circuits will select the display with the most accurate reading automatically.

As with analog meters, protection circuitry is lacking on the less expensive meters. Minimum protection would be a fuse on the current ranges. There are obvious protection advantages to the auto-ranging meters; however, they cannot be expected to know you wish to measure voltage, even though you are still on the resistance scale.

When you select a meter, the best advice is to buy the best meter you can afford, as it will last a long time if properly cared for and not abused. Meters with adequate protective circuits will provide a long life even for the careless operator. If you are experienced in using meters and reading their scales, you may skip the next few paragraphs, and begin reading "Power, or What's in a Watt."

Look at the meter in Figure 1–4. One necessary skill for the experimenter is the ability to read a meter scale. There are really only three steps to reading a meter properly. These are:

1. Determine the range switch setting.
2. Determine the proper scale on the meter face.
3. Interpret the pointer reading

Manufacturers have made it relatively easy to read these multipurpose scales. The range switch setting will determine not only the scale, but the magnitude on the scale. The meter in Figure 1–4 contains four scales; from top to bottom, they are marked on the left side, OHMS, AC, DC, dB. The dB scale is used for comparative AC measurements, and will not be discussed at this time. The AC and DC scales share the same numbers, which are just below the DC scale. There are four rows of numbers, and you must determine, by the setting on the range switch, which row to use. For example, if the range switch is set to the DC V 50 scale, you use the scale marked 0 to 50. That 50 means that the meter cannot indicate more than 50 volts on that range. A voltage of 30 volts would cause the meter to rise to a point covering the 30 on that scale.

Let's look at another example. You have set the range switch to 2.5 volts. The correct scale to use is 0 to 250. There is no. 2.5 volt scale, so you must divide the scale reading by 100. A 1.5 volt battery would indicate to the 150 on the 250 scale. All meters operate in a similar fashion on their voltage and current scales. Only the absolute scales and ranges differ. The most important concept to remember when reading a voltage or current scale is that the number at the right is the maximum you can read. For practice, find the point on the scale in Figure 1–6A (p. 32) where the following voltages will be displayed:

100 volts
25 volts
8.5 volts
0.20 volts

See Figure 1–6B for the correct answers. Remember, always use the range setting that provides the most deflection. Keep the pointer as far to the right as possible.

Reading the ohms scale is slightly different on most multimeters. Notice that 0 is to the right, instead of left. The ohmmeter scale is also nonlinear. That is, the absolute distance between 1 and 2 is about the same as between 20 and 30. This nonlinearity is necessary in order to keep the meter accurate at different levels of resistance. Notice also that the range switch indicates RX1, RX10, etc. For example, if the meter pointer is pointing at 15, and the range switch is on the RX100 scale, the actual resistance is 1500 ohms. Looking at another example, the pointer is pointing toward the 5 and the range switch is set to RX10K. Simply multiply 5 times 10K, to obtain the correct value of resistance, 50,000 ohms. For practice, try to visualize where the pointer would deflect on the

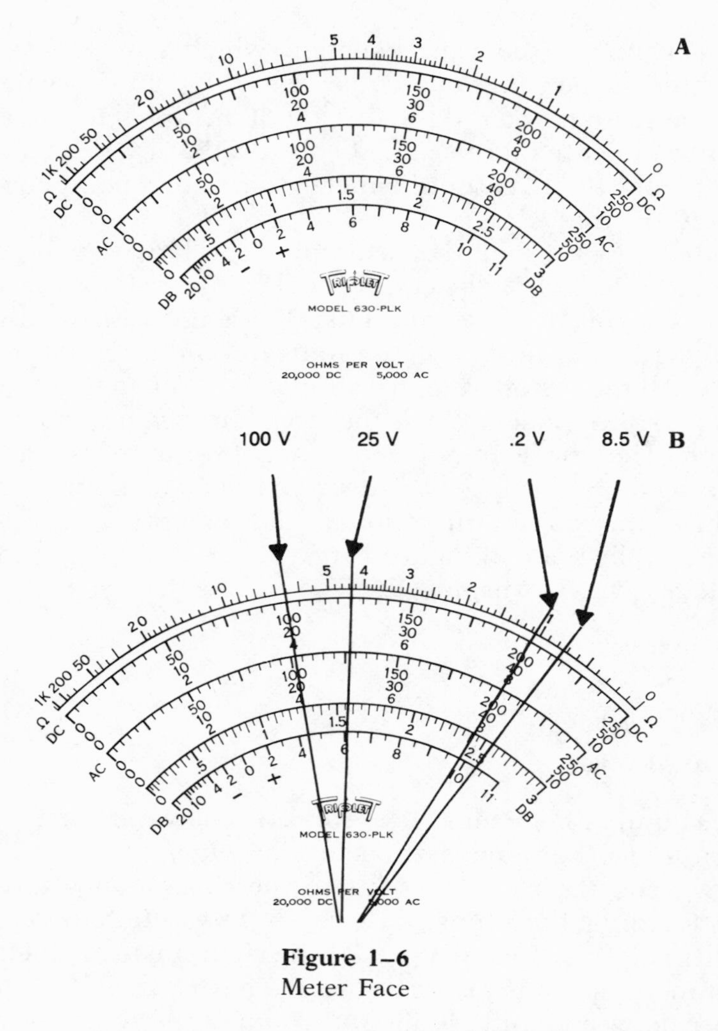

Figure 1–6
Meter Face

following resistances. Use the meter in Figure 1–6A. The correct answers are in Figure 1–7. Resistance:

1000 ohms
750 ohms
25 ohms
45 Kohms

Notice also, the symbol at the far left of the ohmmeter scale. This symbol, ∞, is the symbol for infinity. Infinity means that the resistance of this resistor is so large that the

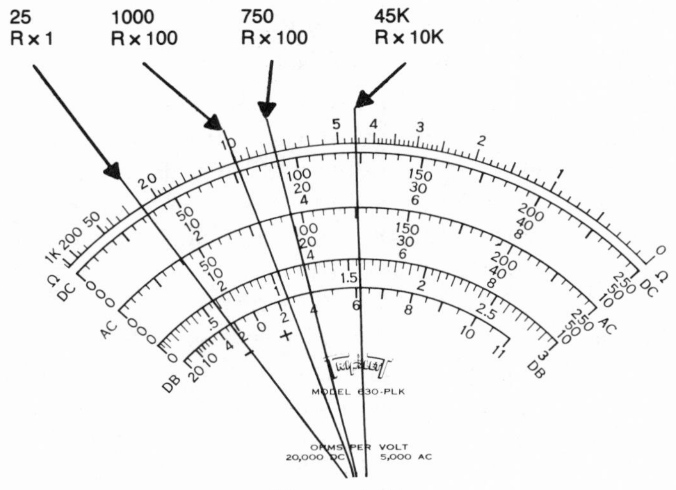

Figure 1–7
Meter Face Solutions

meter cannot measure it. It also can mean that you need to switch to a higher scale.

Unlike reading voltage and current scales, use the reading that deflects the pointer toward the center, as the compressed scale provides the most accurate readings in the center of the scale. Reading a scale takes practice. In the next section you will discover how to measure voltage, current, and resistance in actual circuits. Wire these circuits and measure the parameters with your meter. Do not measure any voltages above 30 volts without first reading the section on electrical safety at the end of this chapter. All of the circuits shown here use a standard 9-volt battery.

To measure voltage properly, there are five steps to follow.

1. Select the correct range switch setting (use the highest scale if approximate total voltage is unknown).
2. Insert meter probes into correct jacks on meter.
3. Check to make sure negative lead is connected to negative point in the circuit. **Note:** Some meters have a +/− switch that allows you to reverse the meter lead hook-up polarity without reversing the meter leads.

4. Apply meter leads to appropriate test points.
5. Apply power to circuit and read voltage on scale.

The simple circuit in Figure 1–8 requires three resistors and a nine-volt battery. If you have a prototype circuit board or soldering equipment, use these to wire the circuit. If you do not, just connect the resistor leads end-to-end, so that the resistors are connected in series. All current must flow from the battery and through all the resistors. See Figure 1–9 for a pictorial of the circuit.

Notice that the battery clip black lead hooks to the negative terminal of the battery, and the red lead hooks to the positive terminal. Notice also, in Figure 1–8, that the shorter line of the battery symbol is connected to R1. This shorter line on schematics is the negative battery pole. Be sure you always check the polarity of the battery hookup when wiring circuits. Some circuits are not critical, however circuits with diodes and transistors will not work, and may be damaged by incorrect battery hookup.

Wire the circuit as shown in Figure 1–9. Use pliers to wrap the leads of the resistors tightly at the ends. These 1K, or 1000 ohm, 1/2 watt resistors and the battery clip are readily available at almost any electronic supply store, or radio-TV repair shop. Be sure to wrap just the 1/4 to 1/2 inch at the ends of the resistor. Wrapping only the ends will allow reuse of the parts in further experiments.

Connect the negative meter lead to the black or minus battery lead. Do not connect the positive meter lead to the circuit until you have completed all steps identified earlier.

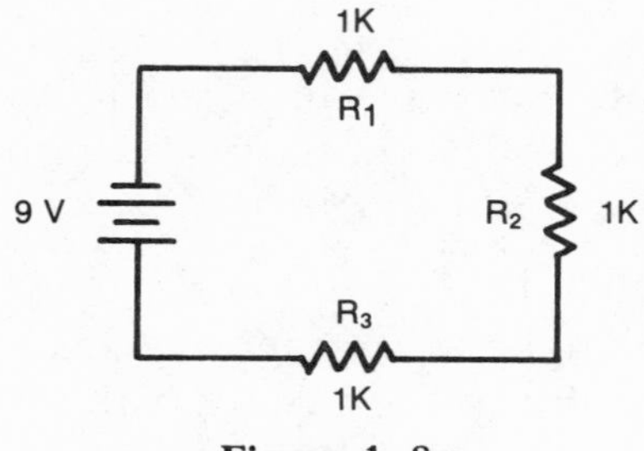

Figure 1–8
Meter Test Circuit

As there is no switch in this circuit, just connect the battery to the snap connector after setting up the meter. The correct meter setting on your particular meter should be on the lowest scale that is larger than the nine-volt battery. Many meters will have a ten-volt scale that would be correct. Apply the meter probe ACROSS R1, as shown in Figure 1–9, Step 1. In other words, connect the positive meter lead to the opposite side of R1 from the minus meter lead. The voltage you read at this point should be approximately three volts. Remove the positive meter lead from the junction of R1 and R2, and connect it across both R1 and R2. In other words, connect it to the junction of R2 and R3. This is shown in Step 3 in Figure 1–9. The voltage at this point should be approximately six volts. After you have verified this, connect the meter to the positive battery clip and measure the positive battery voltage, which is the total voltage in this circuit. This should be approximately nine volts. Notice that you are always measuring across the components. This seems natural, and is the correct procedure for measuring voltage and resistance. When measuring current, however,

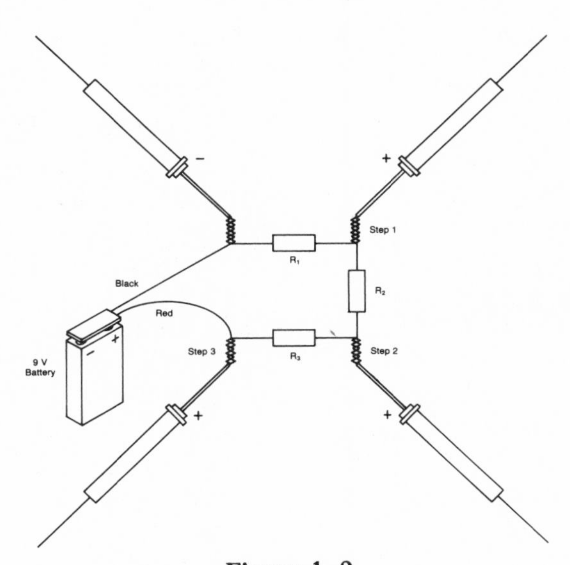

Figure 1–9
Meter Test Circuit Pictorial

the process is different. I make note of it now so that you will be able to more easily relate to this important difference between voltage and current when we measure circuit current.

Measuring resistance is also similar to measuring voltage, but with one important difference. **NEVER** connect an ohmmeter to a circuit that has power applied. An ohmmeter always has an internal battery that supplies a known voltage through the resistor being tested. The meter then measures the current, and displays the resistance reading on the meter, which is calibrated in ohms. If you were to leave the power connected, at best, the reading would be incorrect because of the presence of this voltage, which would cause the wrong current to flow through the meter. At worst, too much current through the ohmmeter, caused by trying to measure resistance with the circuit energized, may cause damage to the internal resistors or the meter movement.

To measure the resistors in our test circuit, follow the steps below.

1. DISCONNECT THE BATTERY.
2. Place the ohmmeter on the R X 1K scale.
3. Zero the meter.
4. Connect the meter across R1. The reading should be approximately 1K.
5. Connect the meter as shown in Figure 1–9, Steps 2 and 3. Step 2 resistance should be approximately 2K, and Step 3 resistance should be approximately 3K.

An important concept to notice about these two sets of measurements that we have just completed is that these components are connected in series, or end-to-end. Voltages in series circuits will always total the supply voltage. This is known as Kirchhoff's voltage law. Resistances in series circuits, like voltages, are additive. In your test circuit, did you notice this as you measured these components?

You have learned to measure resistance and voltage; now let's look at how to measure current. As was stated earlier, when measuring voltage and resistance, you must measure across the component, or circuit. When measuring current however, you must measure current **IN** the circuit. To understand this concept, let's look at what current flow is again. Remember, current flow is actually the flow of electrons in a circuit. Voltage is a pressure applied across a component, so we measure the pressure by measuring across the component. To measure current, however, we must let the electrons flow through the meter circuit. This can only be done by opening up the circuit and inserting the meter leads in the circuit. See Figure 1–10 for a test circuit.

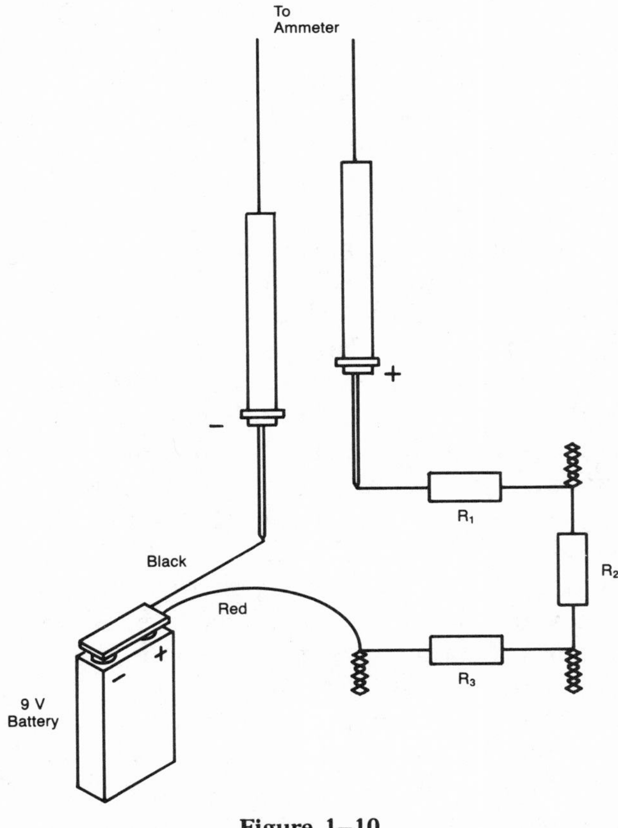

Figure 1–10
Current Test Circuit

Kirchhoff's current law states simply that the number of electrons entering a point will equal the number of electrons leaving that point. This seemingly obvious statement has vast implications. In the series circuit of Figure 1–10, the circuit has been opened at the negative battery terminal. The multimeter should be set on the 10 ma. scale. Remember, when measuring current, as when measuring voltage, be sure to use the largest scale available, when current is unknown. In our circuit, notice that the electrons will leave the negative battery lead and enter the meter lead. After traveling through the meter, they will enter the circuit again, via R1.

We have connected the ammeter in series or **INTO** the circuit, rather than across the circuit. If you wish, you could open the circuit, between R1 and R2, or between R2 and R3. Kirchhoff's current law, however, implies that the number of electrons at these points will be the same. Indeed this is so. Current measured here will be identical to the measurement as shown in the figure. You may measure different current levels by removing R3 from the circuit. Be sure R2 is connected to the battery, and measure the current again at R1. You may also remove R2, and measure the current with only R1 in the circuit. In each case, current will increase as resistors are removed. This is because total resistance decreases, allowing current to rise. Currents will vary between three ma. and nine ma. depending upon the total number of resistors in the circuit, and the condition of the nine-volt battery. You must always measure current in a circuit by connecting the meter **INTO** the circuit, while voltage and resistance are measured **ACROSS** the circuit components. Failure to remember this rule could cause damage to your circuit or your meter. Remember always to open the circuit, and insert the meter at the open circuit points when measuring current.

If you find a used meter for sale, there are a few hints and tests to make to ensure that you have a working meter to take home. When you inspect the meter, look closely at the meter pointer. It should be perfectly straight. Look especially at the lower inch of meter pointer. A slight band near there indicates that the meter received a serious overrange, and the force of meter movement actually bent the needle at the meter stop. Reject a meter like this immediately. It will never be accurate to its original specifications unless a new meter movement is installed.

You may have noticed that some meters have a mirror on their faces. This small mirror runs the length of the meter scale, and is intended to ensure that you look squarely at the meter when reading the scale. Keep the pointer in line with its reflection when taking a reading, to avoid parallax error, which is what happens when your eye is not directly over the pointer, so that the pointer appears to be indicating a value to the left or right of the real value. The mirrored scale will allow you to get the most accurate reading possible from the meter.

Another way to quickly check a meter is to be sure that the meter needle will zero properly. This may indicate trouble, or may just mean a slight adjustment. Be sure you are on the voltage scale, or the meter is in the off position when checking this. A slight misadjustment of the zero can be remedied on some meters by a small, usually plastic, adjustment slot located just below the meter face, underneath the pointer. It adjusts the tension of the front hairspring of the meter movement, and a slight adjustment will usually be all that is required to zero the meter. Be sure to check the ohmmeter zero adjust as well. Remember this is done with the meter leads connected to each other. If you cannot obtain a zero on the ohmmeter, it may mean only that the ohmmeter battery is weak. It may also mean problems in the circuit. Try the ohmmeter test with new batteries before purchasing the meter, and be sure to check each scale for proper zero. If one scale remains open, or will not zero, there could be a damaged internal calibrating resistor. Reject the meter.

These tests will go a long way toward making sure you are getting a working meter, however more operating checks would be in order before making a purchase decision. For a start, bring along the circuit of Figure 1–9 and actually measure the voltages and currents in the circuit, as well as the resistors. Tolerances are additive, so a 20% resistor and a 10% meter that are off in the same direction, might result in a 30% error in reading. This would be no reason to reject an otherwise acceptable meter. If you plan on measuring AC circuit voltages, check the AC line voltage for 120 volts to verify that scale. Also, if you want to check lower scales, obtain a 12 VAC transformer, and measure its output while its input is connected to the AC line. See the chapter on power supplies, and the section on transformers for more details on how to choose a transformer for this test. Also note the safety precautions at the end of this chapter before trying this test.

These simple tests should assure you of getting a good quality working used meter for the cost of a lower quality new meter.

Used meters are sometimes hard to obtain, however here are a few suggestions. Visit any electronics supply house. These places often trade in used equipment, and offer it for sale. Check out any amateur radio operators you may know. These radio hobbyists can locate, among their fraternity, all kinds of test equipment. The radio "hams" have flea markets and conventions where much of this equipment can be found for sale. While you are talking to someone about meters, ask a few questions about the hobby. Ham radio is an excellent entry into the world of electronics.

Another often-overlooked source of used meters is the local TV repair shop. These shops often have meters just sitting around, while the technicians are busy using the new digital meters so often required for the high-technology consumer electronics equipment available today. Many shops would gladly sell an old meter that otherwise might just sit around collecting dust.

Another advantage of purchasing from a professional place such as this is that technicians are professionals, and know how to care for and treat meters properly. Although this is no guarantee of a good meter, it certainly helps. If you select a meter carefully, you will own a piece of equipment that will be useful and durable. It will give you many years of reliable service, requiring only fresh batteries, and an occasional lead repair.

Power, or What's in a Watt

Earlier, the concepts of voltage, current and resistance were covered. The relationships of these three units to each other were mentioned, as well as their individual characteristics. One other concept that should be mentioned, however, is the unit of power. Power is related to the other three units in that it is an actual measurement of energy consumed in the circuit. From light bulbs to motors, most electrical equipment is tagged with its wattage. Most people don't think of wattage in comparing power to energy use. The most usual expression of power that comes to mind is the horsepower. There are 746 watts per horsepower, which should give you an idea of the comparison of wattage to the energy actually consumed.

The relationships of watts to voltage and current is direct. In other words, if the voltage in a circuit rises, then wattage increases also. Similarly, if current increases in a circuit, the wattage also rises. Wattage is inversely related to resistance, however; if resistance goes down, the wattage increases and vice versa. This relationship is explained much more fully in the chapter on math for electronics.

One way of understanding the concept of power is to consider the amount of heat dissipated. The resistors we used in the circuit earlier were specified as 1/2 watt. This specification means that they will dissipate, or transfer into the air, 1/2 watt's worth of heat before they will be damaged by an inability to keep cool. Resistors are available in ranges from under 1/4 watt to over 200 watts, depending upon how much heat is to be dissipated. Many components in electronics are rated as to their wattage, and as an experimenter, you must be sure your circuits do not exceed the capacity of the components.

Some components are designed with large metal surface areas that are to be bolted to a heat sink. A heat sink is simply a large piece of metal, usually finned, that conducts heat away from the component and radiates it into the atmosphere. Many transistors and integrated circuits are given two specifications, power dissipation with a heat sink and power dissipation without a heat sink.

Determining the power handling capability of a component is easy. To calculate the power needed in a circuit, just multiply the voltage times the current. In our resistor circuit, the most power dissipated was 0.08 watts, well under the 1/2 watt resistors specified. The 1/2 watt resistors were chosen because of their availability, even though 1/4 watt resistors would have been adequate.

One other caution regarding power is in order. Components that must dissipate large amounts of power will run hot. Enclosing these components in a cabinet may result in their early demise. Allow freedom of air movement to these components. Also keep exposed skin from contacting power resistors and heat sinks. Temperatures may be high enough to cause severe burns. This brings up other warnings about electrical safety, which we will discuss in the next section.

How to Work with Electricity Safely

There are three hazards common to all electronic shops and work areas. These are electric shock, fire, and expulsion of toxic gases. Each of these will be discussed separately.

People usually think of electric shock when handling electricity. Shock avoidance is necessary when working on any live circuits. The severity of electric shock can range from mildly uncomfortable to fatal. The following techniques will help to ensure enjoyment of your experiments for many years to come.

The severity of an electric shock is dependent upon the duration of the shock, and the magnitude of current flow through body tissue. Fortunately, skin is a relatively poor conductor of electricity, and as a result, voltages under 30 volts usually cannot cause enough pressure to make skin conduction a shock hazard. If the skin is broken, however, resistance is much lower, and severe shock is much more likely. Current must be provided a complete circuit through your body for you to receive a shock. One of the most convenient paths is from one arm across the chest cavity to the other arm. This has prompted the so-called one-hand rule. This rule, stated simply, is that, when measuring voltages above 30 volts, be sure to keep one hand in your pocket, or behind your back. By doing this, there is little chance that it will come in contact with ground, should you happen to contact a voltage point, while you are measuring in the circuit.

Another sensible precaution, when working on high voltage circuits is to stand on a rubber mat. This mat will prevent conduction through the floor, thus minimizing any shock hazard should you contact a live circuit. Ordinary shoes protect to a large degree, in the same fashion, for voltages up to and above 100 volts.

Although most of the circuitry the hobbyist works with is well under 30 volts, there are areas of specific precaution. These include the power supply AC wiring, which is directly connected to the 120-volt lines. Also included are connecting TVs and radios to electronic projects.

Under no circumstances should you connect a project that runs from the AC line to a television or radio that does not contain a power transformer. This means, if you are in doubt, **DON'T**. Many televisions and radios are directly connected from the AC line to the chassis of the receiver. This is of no consequence to the average consumer, as all parts of the set that are exposed are plastic, or glass, or they are insulated in some other manner from the chassis. When you connect a circuit to the receiver, though, you are providing a connection that might leave that set to provide a shock hazard under certain conditions. Proper isolation can be provided by the use of a transformer, however these transformers are expensive, and designed for the television servicer.

Measuring voltages in one of these sets is also hazardous without the use of a transformer to isolate the chassis. *Failure to follow this advice may result in injury or death,* not to mention the possibility of damage to equipment that might occur.

Another hazardous area in power supplies is the filter capacitors. As you will soon discover, the capacitor is capable of storing a voltage for some time after power has been removed. It is a good idea, especially when attempting to measure resistance, to discharge these capacitors after unplugging the supply. This is done by simply connecting a clip lead across the two capacitor terminals. For more on the capacitor, see the next chapter.

An electrical fire is always a possibility in an electronic shop. Trying to extinguish an electrical fire with water, or with an improper extinguisher may be fatal. The electricity from live circuits exposed during the fire will travel down the conductive stream of the extinguishing agent. It may also electrify puddles of the agent that end up on the floor. To avoid this, provide class BC extinguishers, which will not conduct electricity. If you remove the power to an energized circuit, the fire will probably self-extinguish, which could solve the entire problem. Another common safety practice, that also happens to be a real convenience, is to connect all circuits in the shop, or on the workbench, to a single switch. This will allow instant power removal from all components and equipment on the bench in an emergency. It is also quite convenient and saves having to turn off many pieces of equipment when you finish experimenting. Just ask someone who has forgotten to unplug his soldering iron while he went out for dinner.

Toxic gases can also be a problem in the experimenter's shop. The gas is released when components such as electrolytic capacitors become overheated. Provide adequate ventilation for your work area when planning. The smell of rosin drifting through the house while you are soldering that kit together is not very pleasant. More on providing a safe and convenient environment is found in the section on choosing tools and equipment in Chapter 5.

All this is not meant to discourage you from experimenting with electronics. All areas of experimentation contain hazards; just ask the back-yard mechanic. I hope to have made you aware of the greatest hazards, so that they may be avoided. In over fifteen years of electronics experience, of which five have been spent working with complete novices, I have never seen an injury or shock that could in any way be considered serious. This is partly due to the fact that I stress safety above all. It is the most important part of all your experiments. Have fun safely.

A Simplified Guide To Circuit Components

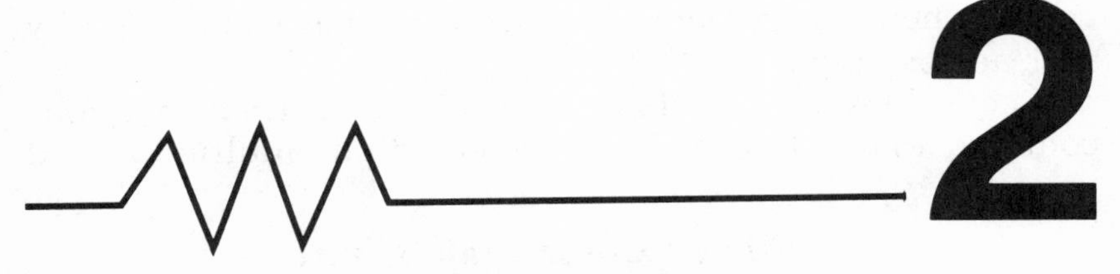

How to Identify and Measure Resistors

In the last chapter you were asked to get three resistors from the local supply store so that you could learn to use the ohmmeter, voltmeter, and ammeter functions of a multimeter, or you purchased them in order to test a used meter you were thinking about buying. If the resistors came in a package, the package was probably already marked with their value. If they were not packaged, did you know their value, or did you rely on the salesman to give you the correct value? You have probably already guessed that their value is expressed in the bands of color that encircle the resistors. Composition resistors are available from 1/8 watt to 2 watts and are color coded. The only exception to this is that certain manufacturers of precision resistors stamp the value of the resistor on each one. You might ask why manufacturers don't do this for all resistors. I am sure it is a matter of economics. The cost of color banding is less than that of stamping the value of each resistor.

Another advantage of banding resistors is that as resistors age, they tend to change slightly in appearance. This is especially true if they are used at or near their rated wattage. By encircling the resistor with color, any changes at one point on the resistor might not affect the opposite side of the resistor. As a result, you may still be able to read the color value long after a stamped number might wear off. The major disadvantage of color coding is in the fact that the experimenter is going to have to either learn the code, or resort to a chart every time he wants a resistor, however the color code is not that difficult to learn. In Table 2–1, the color code is detailed. This chart is a nice reference item to keep handy for use as you are working with circuits. Following the chart are hints on how to read it and how to use it to help you remember the code so that you eventually will no longer need it.

Let's look at the chart in Table 2–1. There are four columns marked first digit, second digit, multiplier and

Resistor Color Code Chart

Color	First Digit	Second Digit	Multiplier	Tolerance
BLACK	0	0	1	
BROWN	1	1	10	1%
RED	2	2	100	2%
ORANGE	3	3	1000	3%
YELLOW	4	4	10000	4%
GREEN	5	5	100000	
BLUE	6	6	1000000	
VIOLET	7	7	10000000	
GRAY	8	8	100000000	
WHITE	9	9	1000000000	
GOLD			.1	5%
SILVER			.01	10%
NO COLOR				20%

Table 2–1
Color Code Chart

tolerance. Each row is marked by the list of colors that are found on the resistors. To use the chart, get a color-coded resistor, or use the resistor drawn in Figure 2–1. Hold the resistor with the color band nearest the edge toward your left. If you try to read the resistor with the other side to your left, you will not get a correct reading. The resistor in Figure

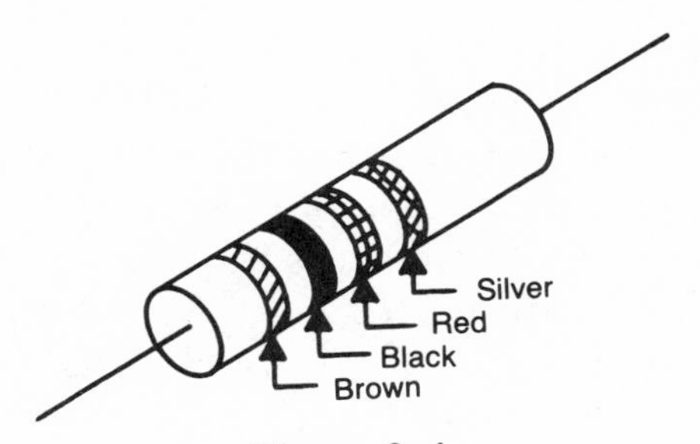

Figure 2–1
Resistor with Color Bands

2–1 is marked with four bands—brown, black, red, and silver. The band closest to the left edge is brown, which corresponds to the first digit. Look in the row marked brown to find the number 1, which is the first digit in the value of the resistor. The second band is black, and referring to the chart, black is 0 in the second digit column. The red third band is the multiplier, which means that the number in this column is to be multiplied by the value of the first two digits.

Let's see how this works, since the first two digits read 10, and the multiplier is 100, we multiply 10 times 100 to get the reading. In this case, the answer is 1000, and we have a 1000-ohm resistor. Another way to interpret the multiplier column is to use the color number as a number of zeros column. In other words, since the third band is red, and red is the digit 2, we can just add two zeros to our reading of the first two digits. In our example, 10 plus two zeros is 1000. This trick works on all numbers for the third band except gold and silver.

The fourth band is tolerance, which is the actual deviation from coded value before the resistor becomes inaccurate. This band is silver on our resistor, which means the resistor has a 10 percent tolerance. The tolerance of this resistor, then, is within 10 percent of its coded value, or between 900 and 1100 ohms. To figure tolerance, multiply the reading of the first three bands by the tolerance band. Add the result to the coded value to determine upper range, and subtract it to determine lower range. If the resistor is in tolerance its measured value will fall within the two extremes. Notice that some resistors have no tolerance band.

No color in the fourth band means that the resistor is a 20 percent tolerance resistor.

Try another example with a resistor marked blue, grey, yellow, gold. The first digit is blue, which is the number 6, the second digit is gray, which is the number 8. The yellow band indicates a multiplier of 10000, or an addition of four zeros to our reading of 68. The total reading then is 680,000 ohms, or 68 Kohms. The gold band is a tolerance of five percent, which means that it can be between 646,000 and 714,000 ohms and still be within tolerance. An ohmmeter indicator that falls within this range will tell you that the resistor is in tolerance.

I mentioned earlier that there is a trick to remembering the order of the color code. As you can see, it is easy to figure out the values if you have a list of the colors in order. Colors, however, are hard to remember, so I learned a technique to remember the first letter of each color. To use this technique, just remember this sentence: "Bad Boys Race Our Young Girls But Violet Generally Wins." The first letter in each words corresponds with the first letter in the color-code list. The hard part is remembering that the list starts with the number 0 rather than 1. Once you learn that sentence, take some time to practice reading resistors using the code. There is no substitute for actual use when you are trying to gain a skill. Open an old radio, or take a handful of resistors and try your hand at reading them without referring to the chart. You will find that it will not take long for you to become adept at reading the resistor values. You can verify the value of the resistors you read by checking them with an ohmmeter.

When measuring resistors with an ohmmeter, be sure to unsolder one resistor lead from the circuit. Resistors, when measured in a circuit, may give an erroneous reading due to the fact that other circuit components are also connected. This rule is good to remember when measuring the resistance of any component.

So far, all we have been talking about are resistors that have a power rating of less than two watts. Most resistors above two watts have their actual value stamped on them. Figure 2–2 contains several resistors of different wattage.

The smallest resistors are 1/4 watt, while the largest is 10 watts. As you can easily see, the higher wattage resistor is physically larger. This is because components that must dissipate extra power as heat must be large physically, so the heat may be more easily radiated into the atmosphere. While you are working on circuits that include high wattage resistors, remember that these resistors run quite warm when in operation and require several minutes to cool down to room temperature. If you accidentally touch one of these resistors, you might receive a severe burn, so be careful.

While on the subject of resistors, we will cover three different types of resistor construction. The small resistors you used in the meter experiments are probably of a carbon composition. These resistors are available in two types, standard and flameproof. Flameproof resistors are similar in construction to the standard resistor, except they have a

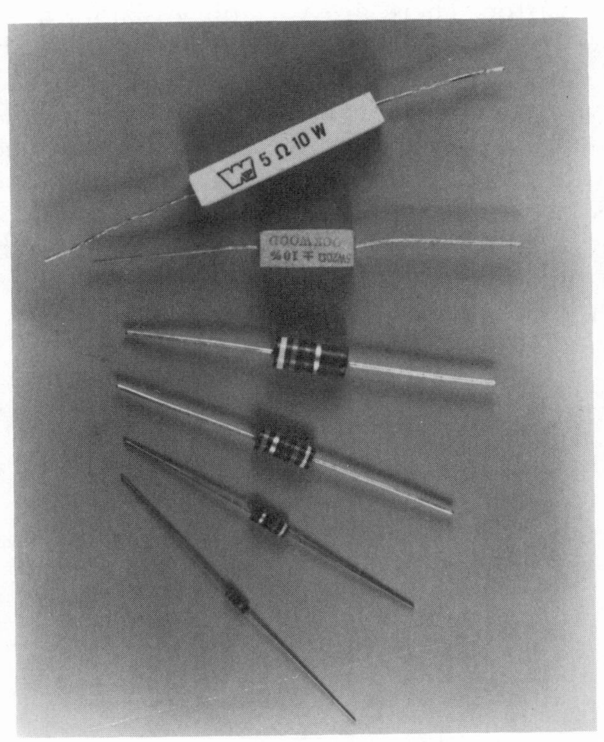

Figure 2–2
Resistors of Several Wattage Ratings

ceramic covering that prevents the resistor from bursting into flame should it overheat. The extra cost of flameproof resistors is minimal, so specify them if they are available.

The resistors in Figure 2–2 that are in long, square blocks, rather than the cylinder shape, are wire wound. Wire-wounds are usually available in sizes from 1 to 225 watts, and are usually found in a ceramic case, which helps them to dissipate heat. Instead of using carbon, these resistors use a high resistance wire that is wound on a form to a certain length. The higher resistance devices contain more turns of wire on the coil.

Another resistor found occasionally is the film type. It is constructed by placing a thin layer of resistive material on a ceramic core. Leads are attached and the assembly is encased in an insulating cover.

The third type of resistor found commonly is the variable resistor. There are variable resistors available in both wire-wound and carbon-track varieties. Variable resistors can have two terminals or three, and are available in many styles of mounting. Different mounting styles allow you to place the resistor directly on a circuit board, or connect it to a front panel, or to a chassis. Different sizes and wattages are also available. Figure 2–3 shows several different variable resistors, from miniature printed circuit mount trimpots to high wattage chassis mount controls. The term "trimpot" stands for trimming potentiometer.

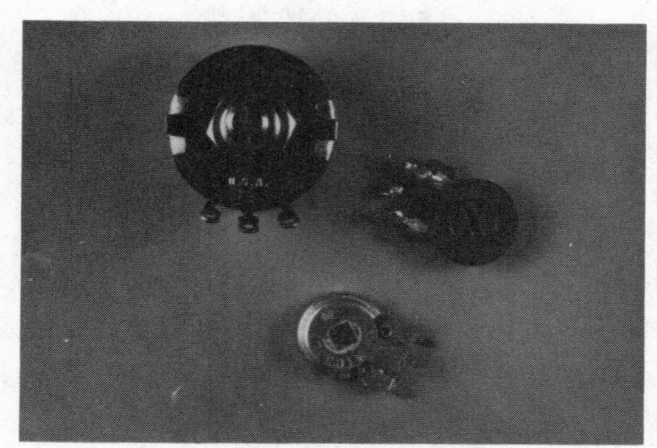

Figure 2–3
Controls and Trimpots

A potentiometer is the correct name for a three-terminal variable resistor while a two-terminal device is called a rheostat. Figure 2–4 shows the schematic for the potentiometer and the rheostat. The potentiometer can be used as a rheostat by connecting one outer terminal to the middle, or variable terminal, while a rheostat usually cannot be connected like a potentiometer, as there is no third terminal. As you can see by the schematic, the variable resistor is a simple device. If you have an old one around, take it apart and look at its construction. A carbon track is connected between the outer terminals, and a metal wiper moves along the track as the control shaft is adjusted. Use an ohmmeter and measure the two outside terminals. These terminals have the maximum control resistance across them, and you order a potentiometer by that value. In other words, if you measure 10,000 ohms across the two outside terminals, it is a 10,000-ohm, or 10-Kohm pot.

Now measure between the center wiper terminal and either end terminal. You will find that as you turn the shaft, the actual resistance will vary. Turn the control to one end stop, and you will measure near zero ohms between the wiper and one terminal, and maximum resistance between the wiper and the other terminal. Ideally, the minimum resistance should be zero ohms, however all composition variable resistors have a value slightly higher than zero because of the "hop-off" resistance of the wiper assembly.

The carbon-track variable is available in low wattage styles and wattage rating is based on using the full resistance element, however higher power variables are also

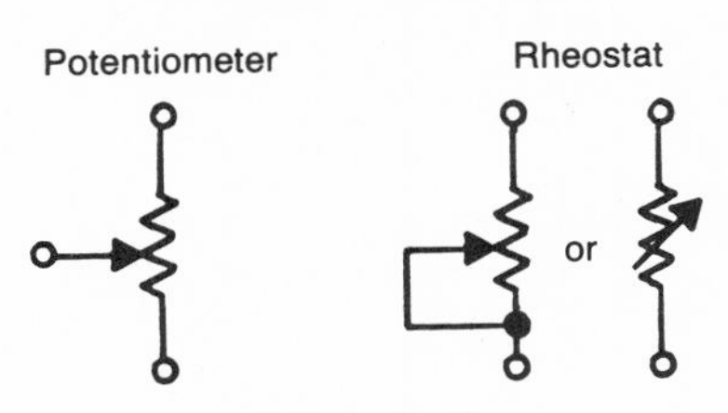

Figure 2–4
Potentiometer and Rheostat Schematic

available. These variables, like their high wattage counterparts are usually wire wound. The only difference in internal construction is in the fact that the carbon track is replaced by an insulated core that is wrapped with high resistance wire. The wiper is a metal piece that comes into contact with the wire edge and selects the resistance in exactly the same manner as the carbon-track variety.

How Capacitors Work

One of the most interesting devices is the capacitor, and it is quite easy to understand and use. The capacitor has three discrete functions, which are:

1. The ability to store energy.
2. The ability to block the flow of DC.
3. The ability to allow AC to flow across the device.

Each of these functions will be covered in order, after an explanation of the construction of the capacitor.

Referring to Figure 2–5, look at the drawing of a basic capacitor. Simply stated, the device is two conductors separated by an insulator. The insulator can be anything that normally is classed as an insulator; however, the most commonly used materials are paper and plastic film. There

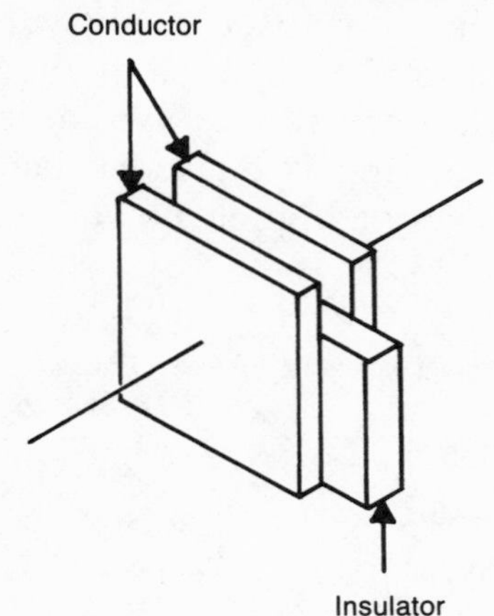

Figure 2–5
Drawing of Capacitor

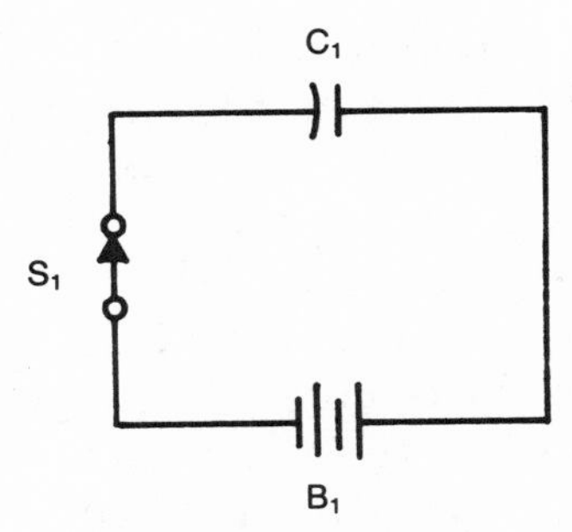

Figure 2–6
Capacitor Circuit

are many other insulator types, which will be discussed in the next section.

Conductors are usually aluminum foil or, in the case of variable capacitors, aluminum plates. To see how they work, look at Figure 2–6, which contains a battery, a switch, and a capacitor. Notice that the schematic symbol of a capacitor is almost like that of a battery. You can tell them apart by the fact that the battery has uneven lines while the capacitor has even parallel lines. To understand how the capacitor works, keep in mind that the effect the capacitor has on a circuit varies with time. This means that we will have to analyze the operation of the capacitor over a period of time.

At the instant the switch is closed, a completely discharged capacitor will look like a direct short to a battery. That's right—even though there is no direct electrical connection to the other side of the battery, it allows current to flow heavily. With a large battery and a large capacitor, this might be several amperes of current flow. This large current flow, however, only lasts for an instant as the voltage begins to rise across the capacitor. Since the voltage across the capacitor is low to begin with, electrons are attracted to the plate. These electrons continue to build on the plate until their negative charge repels the addition of further electrons on the plate. The electrons are attracted through the insulator toward the positive plate on the other side. The capacitor is now charged. Even if you were to remove the battery, the voltage would remain on the plates for some time. The length of time it will remain there is dependent upon the size of the plates and the thickness and construction of the insulator. Capacitors thus are capable of storing energy, which has a multitude of useful applications in electronics.

Notice that once the capacitor is fully charged, no more current will flow in the circuit. The capacitor now looks like an open circuit, rather than the short circuit it imitated earlier. We have seen that the capacitor can store DC and block DC from flowing through the circuit, once the capacitor is charged. Now let's see how the capacitor allows AC to flow. AC does not actually flow through the capacitor, it literally flows *around* it. As the voltage in the AC circuit starts to rise, the capacitor is discharged, and electrons start to gather on the negative plate. Before the capacitor can be completely charged, however, the voltage starts to return to zero. The capacitor now starts to discharge in the direction of the AC voltage. This discharge tends to oppose the lowering voltage, thus causing the AC to seem to delay its return to zero. Upon going negative, the other capacitor plate begins to collect electrons, allowing the capacitor to charge to the opposite polarity. Upon completion of the AC cycle, the capacitor is now charged in the opposite direction, and the next AC cycle begins.

As you can see, the electrons travel around the capacitor, through the load and source, then arrive essentially at the other side. This gives the effect of allowing AC to pass through the capacitor. It is interesting that this effect is amplified as the frequency of the AC is increased. This will be explained further soon. You will find capacitors extremely useful when you want to select a certain frequency, or group of frequencies, and reject others. The capacitor, when used in combination with a coil, brings us the miracle of radio communications.

Capacitor Types and How to Identify Them

As was mentioned earlier, there are several classes of capacitor. Capacitors are sorted by insulator type and actual value. Up to now we haven't looked at a capacitor's value, so let us do that now. The single unit of capacitance is the farad, which is derived from the name of an early experimenter named Faraday. One farad represents the amount of charge on the capacitor if it stores one coulomb of electrons with one volt on its plates. For nearly all circuits in general use, one farad is much too large. Indeed, most capacitors are rated in microfarads. The prefix "micro" means millionths, in other words, one microfarad means one millionth of a farad. A microfarad is even too large for many small capacitors, which are often rated in nanofarads (.000000001 farads), or even in picofarads (.000000000001 farads). Another term you may hear is "micro microfarads," an obsolete term that means the equivalent of picofarads.

The actual value of capacitance is dependent upon three characteristics, which are:

1. The total plate area.
2. The distance between the plates.
3. The insulating material between the plates.

By changing any of the above criteria, you will change the value of a capacitor. Figure 2–7 shows three capacitors, each with the same value of capacitance. They are all different sizes, even though their values are the same, because they have different plate sizes, spacing, and insulating materials. The one on the left is a disk ceramic type, the center capacitor is a mylar, and the right one is an electrolytic. When specifying capacitors, you must identify the

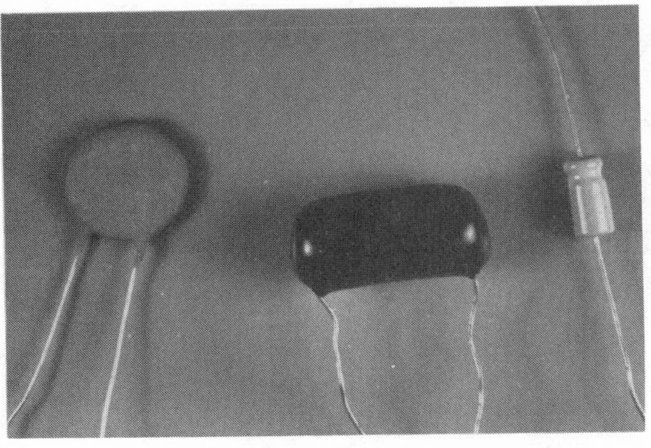

Figure 2–7
Capacitors, Ceramic, Mylar and Electrolytic

insulator, as well as the value. Tolerance varies with insulator value as well as quality. In other words, the percentage of tolerance varies with the insulator material, as well as the grade of material used. Certain materials are more stable than others, and as a result, have more consistent readings. This may be an important consideration if you are building critical-timing, or frequency-generating circuits. In this case, buy high quality temperature-stable capacitors. Capacitors that are not temperature stable will actually change value with a change in operating temperature. Most circuits do not need highly stable capacitors, and all projects in this book will specify a higher grade of capacitor stability only when needed.

Table 2–2 contains a list of typical insulator materials. Most experimental projects will perform adequately with

Capacitor Insulator Types

AIR
CERAMIC
ELECTROLYTIC
MYLAR
PAPER
VACUUM

Table 2–2
Typical Capacitor Insulators

mylar, or ceramic types, but capacitors above 0.1 mfd. will probably have to be electrolytic. This is because electrolytics provide a large capacity in a small package. Let's see why.

Electrolytic capacitors are specially constructed with rolls of foil, one of which has an impregnated oxide coating on one side of the foil. The oxide, though very thin, is an insulator. The other plate is made of gauze soaked in a chemical paste. The paste is called an electrolyte, a chemically active agent that has the property of acting as the other plate. The other foil is not a plate, and only provides a connection to the electrolyte. Unlike other insulators such as mica, or mylar, the thin oxide layer insulates only when the electrons are on the electrolyte. If the electrons are allowed to flow to the other plate, the insulating characteristic is broken down, and the capacitor becomes a conductor. As a result, this type of capacitor is marked with polarity indicators. There is a + or a − symbol on the appropriate lead of the capacitor.

Caution must be exercised when using electrolytic types, as they must be installed in a circuit correctly. They cannot be used in AC circuits where the voltage changes polarity, though they will work in AC circuits that change voltage without changing polarity. The consequences for incorrectly connecting an electrolytic capacitor are severe. Since it becomes a conductor when hooked up backwards, current starts to flow across the insulator. This current causes the capacitor to heat up, which in turn, causes the paste to expand with rising temperature. Eventually the capacitor cannot contain the expanding paste, whereupon it explodes, throwing bits of metal, paper, paste and smoke all over the general area. This is a very real hazard if you are working on an exposed circuit, as it would be easy to get bits of blown capacitor in the face and eyes. Be **SURE** that the polarity of an electrolytic is correct by checking the circuit with a voltmeter **BEFORE** installing the capacitor. **NEVER** connect an electrolytic capacitor to an AC circuit.

Another characteristic of capacitance is capacitive reactance. This property is related to the capacitor's effect on AC voltages. It will be explained more thoroughly in Chapter 6.

The last parameter needing mention here is the capacitor's voltage rating. Capacitors are specified according to the voltage expected across them. Most disk capacitors used to be rated to 600 volts, however solid-state circuits have made lower voltage ratings more common. This means that a circuit with 25 volts across it can use a capacitor, as long as its voltage rating is larger than 25 volts.

How Coils and Transformers Work

In the last chapter, it was pointed out that the transformer was capable of changing voltage and current levels as needed. In this section, we will explore this characteristic, as well as the functions and characteristics of the coil. Before we begin, we should spend a little time talking about the property of electricity that is responsible for the phenomenon. As was stated earlier, there is a magnetic field that surrounds a conductor when it is carrying current. Figure 2–8 demonstrates what this field might look like if you could see it. This electromagnetic radiation is the domain of radio waves, and is caused by varying voltages and currents in the wire. In other words, it is the property of alternating current circuits; as was pointed out earlier, a transformer will not work on DC.

The lines of force radiating from a wire, when it is close to another wire, will actually cause a voltage to be developed in the other conductor. This is called induction because the voltage is induced in the wire. Coils of wire carrying AC voltages actually are capable of self-induction; in other words, one turn of wire in a coil induces a voltage in all the nearby turns. Each turn affects the next turn, all through the coil. This self-induction causes the coil to oppose the AC voltage applied to it. What this means is that, in AC circuits, there is a new type of opposition to current flow, similar to resistance. This resistance to AC is called reactance, and it is so similar to resistance in AC circuits that it is measured in ohms. To keep these two terms straight, think of reactance as AC resistance. Reactance will be explained in greater detail in Chapter 6. The reactance of a coil, unlike resistance, cannot be stamped on the side of the coil because of one special effect. As the frequency of an AC waveform increases, the reactance also increases. The opposition to AC actually increases due to the more rapidly changing AC waveform.

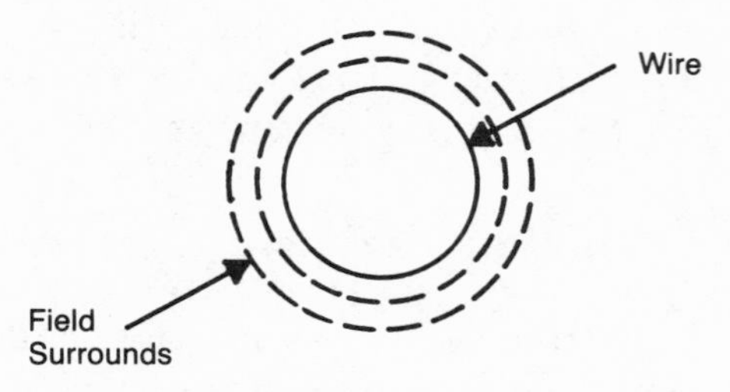

Figure 2–8
Magnetic Field Around Coil

If we cannot rate a coil by its reactance, what can we use? There is a property of a coil that doesn't change with frequency, called inductance—the ability of a coil to self-induce a voltage in its windings.

The unit for inductance is the henry, named after an early pioneer in the field. As with other values, the metric prefixes, milli and micro are used to denote small value coils. The properties of a given coil and its effects on an AC circuit can be easily calculated, and the chapters on electronics math and AC circuits explain this further.

There are two styles of coil in common use, the air core coil, and the iron core coil. Figure 2–9 shows the schematic diagram for both types, and the black lines next to the iron core coil simulate the core. The illustration also contains the schematic representations of the adjustable coil and powdered iron core coil. The effects of induction are actually magnified when the coil is wrapped around a metal core, so large value coils usually have an iron core. Air core coils are used when only small values of inductance are required. Figure 2–10 is a photograph of several coils of different types. As a general rule, the larger the coil, the more reactance it will have at a given frequency. Air core coils have two types of forms available. Fine coils made of small wire that are to be air core types are usually mounted on a cardboard coil form that has no effect on the value of the coil and is there only to hold the coil up. Coils wound with heavier gauge wire are self-supporting, and need no cardboard forms to support them. When placing coils in a circuit, the experimenter must be careful not to put a coil in

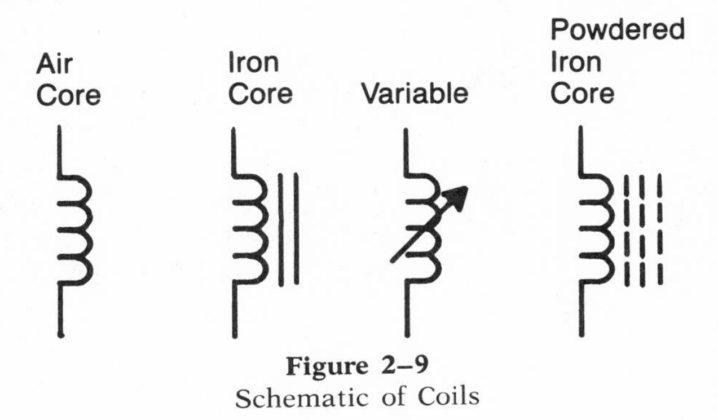

Figure 2–9
Schematic of Coils

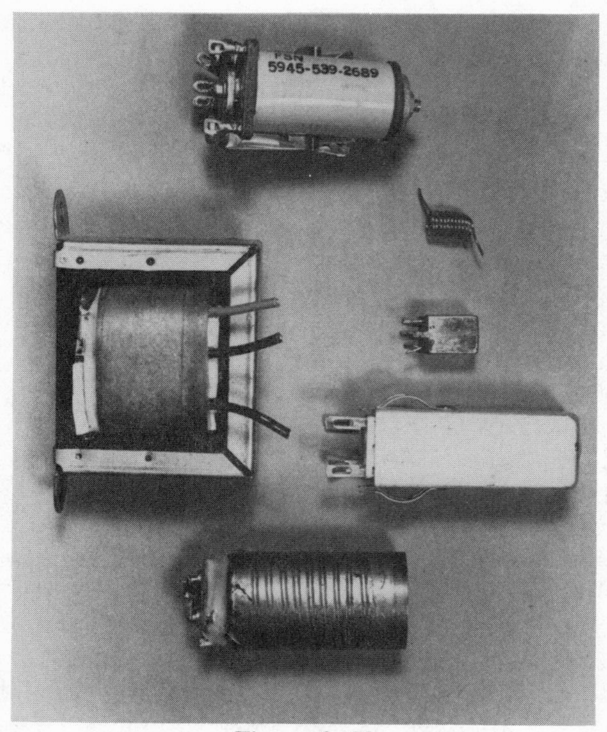

Figure 2–10
Several Different Styles of Coils

proximity to another coil, as one coil will cause a voltage to be induced in the other coil. If coils must be placed nearby, they can be mounted at right angles to each other, which will minimize the possibility of induction. If this technique doesn't provide enough isolation, and the circuit is critical, or doesn't work properly, the next step is to shield the coils with a metal case. This is a common technique; just open the back of any small radio and look. Some of those small metal squares with the screwdriver slots are actually coils. The screw slot is a powdered iron core, and by varying the position of the core, you will be changing the inductance of the coil and affecting its reactance to certain frequencies. Never adjust these coils without the proper test equipment as you will ruin the radio's capacity to receive stations properly.

The action of induction, we saw, could be transferred from one coil to another. We can put this to work for us, as we found out in Chapter 1 when we looked at the transformer. The induction from one coil toward another is improved as coils are placed closer together, and, as a result, transformers are usually wound with one layer of wire wrapped over the top of another layer. This provides maximum transfer of energy. Transfer of energy is also improved when the transformer has a metal core, so laminations of iron are often used to help concentrate the energy in the core. Laminations are used because a solid core is not as efficient. Each lamination is coated with an insulating material, so there is no electrical conduction between iron parts of the transformer.

All of this seems like a "free lunch," and we all know there are no more free lunches, so let's check this further. Figure 2–11 shows several transformer styles. There are air core and iron core transformers, each style used for different frequencies. A low frequency needs a large coil and a metal core in order to efficiently transfer energy from one winding to the other. Figure 2–12 shows the schematic diagram for both transformers. The black lines in the center of the two windings represent an iron core, just as in the schematic of a coil. The winding that inputs energy, such as 120 volts AC from the wall outlet, is called the primary winding, and the other winding is the secondary. Energy transfer from the primary to the secondary is related to the number of turns in each winding. If you were to build a transformer with 100 turns on the primary, and 10 turns on the secondary, there would be a 10:1 voltage ratio. In other words, 120 volts in would give 12 volts out of the secondary. If the circuit on the output of the secondary of the transformer required two amps of current, the primary would deliver slightly more than 0.2 amps. Power going into a circuit must equal power leaving a circuit. In the last chapter, we discovered that power was voltage multiplied by current, therefore power output by the transformer is 24 watts here, while input power is also 24 watts. Notice that I said "slightly more than 0.2 amps." The extra current is being delivered to the

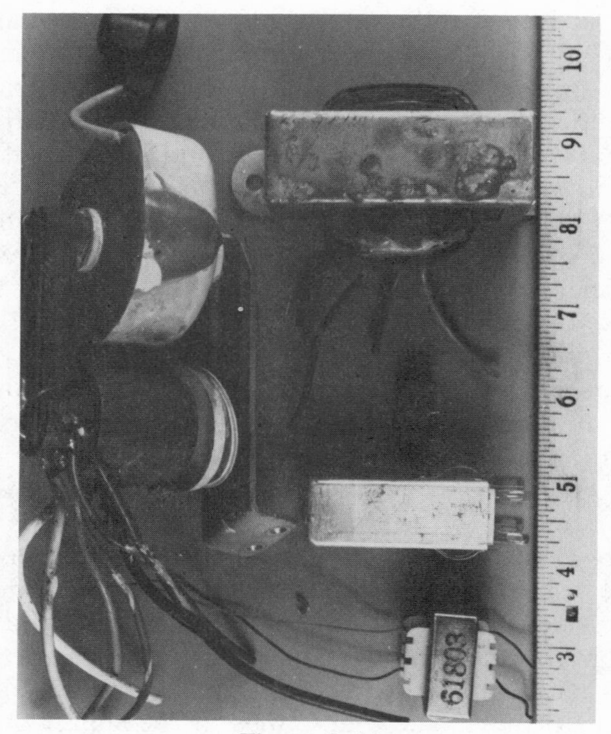

Figure 2–11
An Assortment of Transformers

transformer, which uses it as heat generated by losses in the transformer. A real world transformer is usually between 80 and 90 percent efficient—power output will be that much less than power input.

A transformer that delivers less voltage from its secondary than its primary input voltage is called a step-down transformer. If we can step down voltage with this transformer, we can also step up voltage. All that has to be done is to increase the number of turns on the secondary to more turns than are on the primary. A transformer with a 1:10 turns ratio will deliver 1200 volts to a load, if the primary voltage is 120. If our primary is delivering 0.2 amps of current or 24 watts, though, the secondary will be able to deliver only 0.02 amperes, again 24 watts. Actually, it will be slightly less than 24 watts, due to losses in the transformer.

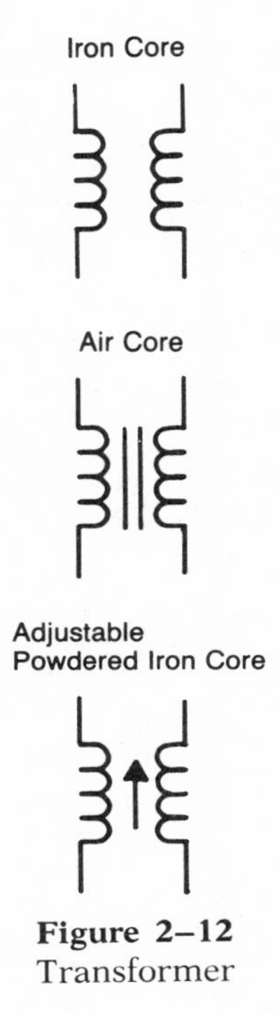

Figure 2–12
Transformer

It is possible to have more than one secondary on a transformer, with each winding delivering a different voltage. It is also possible to put a "tap," or extra output wire, somewhere in the middle of a winding. These taps and extra windings, when asked to deliver current, will demand extra current from the primary. Remember, the sum of all secondary currents will be slightly smaller than the primary current. More information on transformers, especially for selection in power supplies, can be found in Chapter 7.

How to Test Coils, Transformers, and Capacitors

We've spent a lot of time talking about how these devices work. Now let's see how they can be checked. An ohmmeter can be used to test these parts very easily. A coil is simply a piece of wire wrapped around a form, right? Well, an ohmmeter can check to see if there is an open wire in the coil. Most coils should have less than a few hundred ohms of resistance. Coils that are in metal cans should be checked to see if there is any connection between the wire in the coil and the can. If there is any indication at all that there is any less than infinite resistance, the coil is probably bad.

Transformers are checked just like coils, though it is sometimes difficult to find out which wires are to be connected, and which aren't. Some transformer leads are color coded, so that the windings that are connected have low resistance, while wires of different colors should have no connection between them. A tapped winding is usually striped with one of the colors the same as the solid color winding that it is tapped to. See Figure 2–13 for an example of a tapped transformer. If there are no colors on the windings, you will have a difficult time matching windings. The only solution to this problem is to use an ohmmeter to measure the windings, while looking for pairs. Any odd number combination will be a winding with a tap, and resistance to the tap will be lower than the resistance across both ends. There should be no connections to the metal part of the transformer core from any wires. If there are, the transformer windings are shorted to the core, and the transformer is defective. If you know the input and output voltages, and which winding is connected to each, you could test them by connecting the primary to an AC voltage source, and measure the secondary with an AC voltmeter. *Be careful*—if you were to accidentally hook the 12-volt secondary of a transformer to the 120-volt line, you would find that the primary would deliver about 1200 volts. This is a hazard,

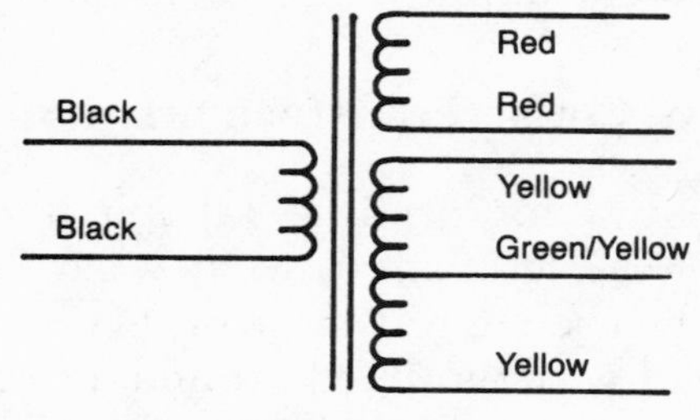

Figure 2–13
Center-Tapped Transformer

not only because of the high voltage, but because most transformers have insulation rated only to 600 volts. Need I say more? If you test a transformer dynamically, use a low voltage AC, such as 12 volts on the primary, and measure the secondary with the meter on the highest setting to start with. Multiply any reading by 10, and you will have the unloaded voltage rating of the transformer. Since there is no current flowing (aside from the small meter current), the voltage is higher than it will be in a working circuit. Using these rules, you should be able to test just about any coil or transformer for continuity.

After seeing how easy it is to check coils and transformers, what about capacitors? Checking these devices is easy, and they can be checked with an ordinary ohmmeter. We have already said that capacitors will pass AC and block DC. A capacitor should check open if this statement is true. The meter tells us more than that, though, if we watch it carefully. Let's take a large value capacitor and connect it to your ohmmeter. Make sure the capacitor is discharged, set the ohmmeter scale to RX 1 megohm, or its highest scale, and watch the needle as you connect it to the capacitor. Did you see the meter needle rise, and then fall back to infinite? Discharge the capacitor by connecting its leads momentarily, then try it again, and watch closely. If you used a capacitor larger than 1 microfarad, you probably saw this deflection easily. Smaller capacitors will move the meter only slightly, and capacitors smaller than 0.001 mfd. are very difficult to check. What you are seeing is the capacitor charging to the ohmmeter battery voltage. Once a capacitor is charged, it represents an open circuit to a DC voltage. You can use this process to test a suspected defective capacitor by testing a known good capacitor, and comparing the amount of meter needle movement to the suspected device. A capacitor with no movement at all indicates an open capacitor, while any value of resistance that remains after several seconds indicates a leaky or shorted device. Either way, replace it. One caution, after the capacitor is checked once, it will be charged. If you try to repeat the test, it will immediately indicate open unless you first discharge it by connecting its two leads to each other. Very small capacitors store so few electrons that this test is invalid, as they charge so fast that the needle doesn't have time to move. For these, you must use a capacitor checker. Most experimenters live without these as they are really too expensive for the occasional user. Take your capacitor to a radio or TV service agency and ask them to check it for you. Don't be surprised if they charge you for the test. After all, you don't have money invested in a checker.

Diodes–What They Are and How to Identify Them

In Chapter 1, you learned about semiconductors, elements that are better conductors than insulators, but not as good as conductors. The semiconductor is the heart of the solid-state age. One of the earliest solid-state devices is the semiconductor diode. Let's take a look at one and see what it is composed of. The diode is made of specially treated semiconductor materials. Common materials used in diodes today are silicon and germanium, though selenium is sometimes used. The silicon diode is most often found in power handling and switching applications, while germanium is usually found in smaller current applications.

The special treatment these materials receive is essentially the addition of impurities into the atomic structure of the material. These impurities cause the semiconductor to contain either more or fewer electrons than a normal piece of material. A normal piece of silicon, with impurities that make the material have a surplus of electrons, is called N-type silicon, while silicon that lacks electrons is called P-type. The P-type material has holes, or spaces where electrons should be, but aren't. Many books go into great detail about the concept of hole flow in semiconductors; however, simply stated, when an electron leaves an orbit, the space it leaves becomes a hole. If you could see electrons moving from left to right in a circuit, you would also be able to see these holes moving in the opposite direction—this explains the process of hole flow. It should be pointed out that these devices are not batteries, or voltage sources that are capable of delivering a potential. It simply means that

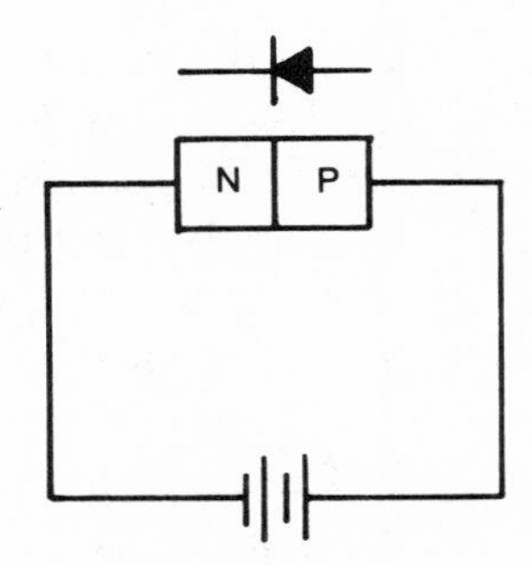

Figure 2–14
Diagram of Junction Diode

the impure silicon contains more or fewer electrons in its orbits than a normal piece of silicon. Figure 2–14 shows a diagram of two pieces of silicon, one with P-type impurities and one with N-type. This device is called a junction diode, which means it is two pieces of silicon, both N-type and P-type, connected at a junction. It is connected to a battery with its N side toward the negative lead. The surplus of electrons on the negative battery pole literally pushes the excess electrons across the junction, where they recombine with atoms that are lacking electrons on the P side. Though it looks as if the holes might be filled, the potential of the positive battery pole keeps electrons moving through the material, causing a continuous current flow in the circuit.

The symbol in Figure 2–14 above the P-N blocks is the schematic of the diode. The arrow of the diode is called the anode, while the vertical bar at the left is called the cathode. The cathode of a diode is always the N-type material. Figure 2–15 is a schematic of a diode with the cathode connected

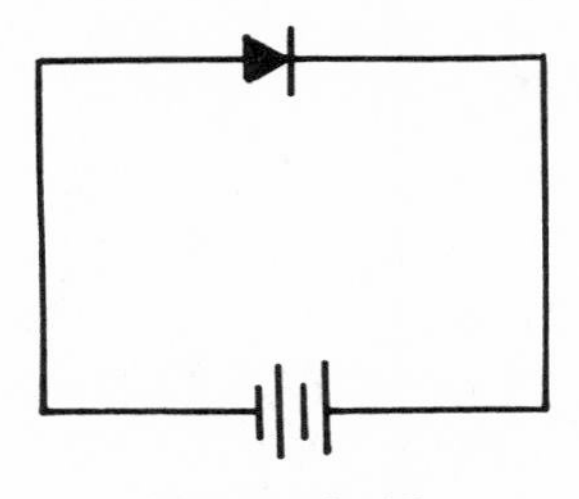

Figure 2–15
Reverse Biased Diode

to the positive battery terminal. In this hookup, the diode's N material is connected to positive, and the electrons will be attracted toward the battery. The negative battery terminal pulls the holes toward itself, thus very little, if any, current flows when the diode is connected in this manner. As you can see, by reversing the polarity of the voltage, you can alternately turn the diode on and off. Using this principle, experimenters have found many uses for this versatile device. Although this is a silicon diode, all of this description applies to germanium diodes as well. The practical difference between these two devices is that the silicon device is usually capable of handling higher voltages and currents.

Chapter 7 details the application of a diode in its most used circuit, the power supply. By using the automatic switching ability of the diode, you can easily change the AC line voltage to DC. To use a diode, you need a few simple guidelines to help you select one for a particular application. A diode must be able to handle all the current you send through it when you switch it on, so check a diode's forward current rating for this specification. This specification is usually called Ifwd, or some variation, in the semiconductor manuals. For example, a conducting diode that is forward biased with a current of 0.5 amperes should be at least one ampere forward current rating, so that it doesn't overheat and burn out. "Forward bias" is the term experimenters use to indicate a diode with a negative voltage on its cathode and a positive voltage on its anode, as drawn in Figure 2–14. A diode, when used as a switch, should be selected so that its reverse bias voltage is less than the PIV rating of the diode. Reverse bias is shown in Figure 2–15, and PIV stands for peak inverse voltage, the reverse conducting point of the device. Choose a PIV that is at least twice as high as the applied voltage on the diode.

Chapter 9 has more details on replacing diodes in existing circuits, but let's look at how to test a diode to see if it is good. Here is another application for our trusty ohmmeter. Figure 2–16 is a photograph of several diodes in different case styles. The actual size of a diode will vary with its current capacity, the smallest diodes pictured having one ampere current capacity. Notice that most diodes have a

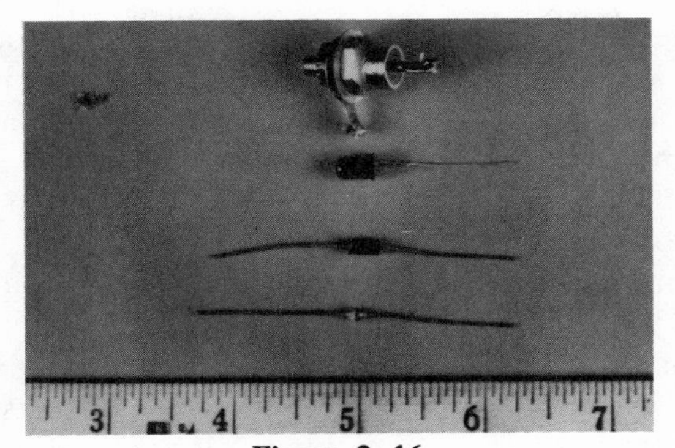

Figure 2–16
Diodes: Top to Bottom:
Stud Mount, Light Emitting, Silicon Rectifiers

band or some other marking on one end. This banded end is the cathode, and, you will remember, this end should be the N-type material. By removing at least one end of the diode from its circuit, you can easily test it. If you put the negative ohmmeter lead on this end of the diode, it should conduct, causing a low resistance indication on the meter. If you reverse the ohmmeter leads, putting the negative lead on the unbanded end, the diode will be reverse biased. The ohmmeter indication should be high resistance. Figure 2–17 illustrates this process. You may notice that by changing ohmmeter ranges around, the actual value of resistance varies, both forward and reverse biased. This is because of an interesting characteristic of a diode, explained in the following paragraph. Keep this in mind when measuring diodes, and keep the meter on the same range for both

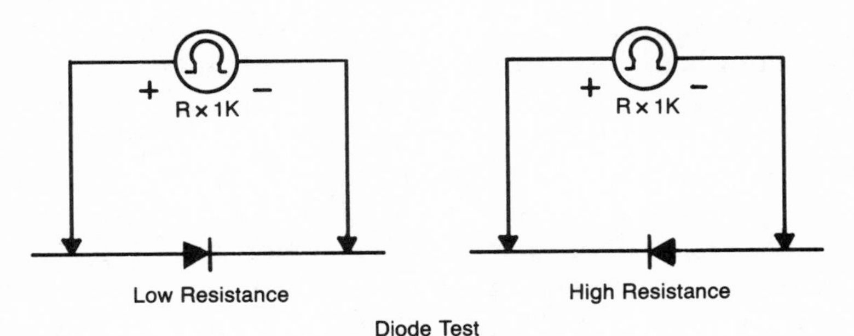

Figure 2–17
Diode Test Hookup

measurements. A good diode will measure ten times or more resistance in reverse bias than it did in forward bias. The actual value will also depend upon whether it is a silicon or germanium device. Silicon devices always have a ratio far greater than 10 to 1 if they are good; 100 to 1 is more common for them.

The interesting characteristic involving the diode concerns the junction of N-type and P-type materials. This junction, as stated, acts as a barrier in the reverse bias mode that keeps the diode from conducting unless this potential barrier is broken down. By increasing the potential, the barrier is broken down, and the diode conducts. There is also a barrier in the forward biased mode, however it is more easily broken down. In fact, an ordinary ohmmeter battery will provide enough potential to overcome the barrier and allow the diode to conduct. With a silicon diode, this potential is fixed at 0.6 volts, and a germanium diode is fixed at 0.2 volts. A single 1.5 volt battery allows the diode to conduct with no trouble at all. This is what we are doing with the ohmmeter check. The ohmmeter battery causes the diode to conduct when it is forward biased, thus breaking down the potential barrier.

There are diodes that have other applications besides switching. These include variable capacity diodes, tunnel diodes, and zener diodes, among others. The most common diode next to the switching diode is the zener diode. Zener diodes are special diodes designed to operate in the reversed biased mode. These diodes will not check properly in circuit, as they are not being operated as forward biased diodes. The zener diode is often found in power supplies, and maintains a constant voltage output, within limits, even if given a varying voltage input. These devices look almost identical to the regular diode (see Figure 2–18), except that they are connected with their banded lead toward the positive terminal. As a result, this constant output characteristic is used to reference, or compare a zener's output to that of an external circuit. See the regulator circuits in Chapter 7 for an example of zener reference circuits. We are just begin-

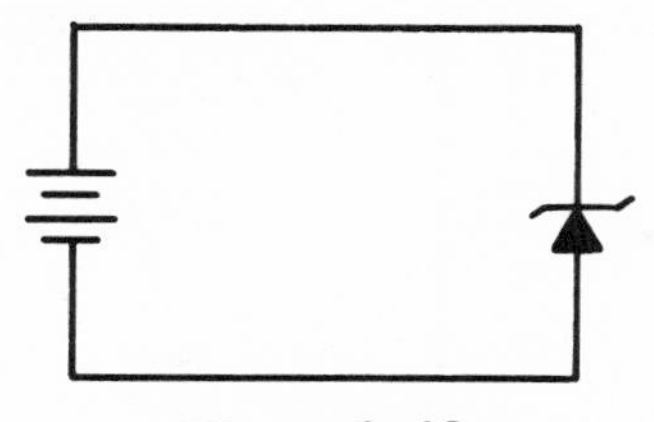

Figure 2–18
Diagram of Zener Diode

ning to delve into the mysteries of the semiconductor. Without it, this marvelous electronic age would not exist. The next section discusses the component that did more to revolutionize electronics, and indeed industrial technology, than probably any other invention, the transistor.

Transistors—How to Identify and Test Them

When Shockley, Bardeen, and Brattain were looking for a solid-state switching circuit to replace vacuum tubes in Bell telephone circuits, they looked to the special properties of the semiconductor. With the discovery of the transistor in 1948, a new age was born, an age of miniaturization, an era of technology never before seen.

The bipolar transistor is, in many ways similar to the junction diode just discussed. A bipolar transistor is the most commonly used transistor in circuits today, though there is another kind. The field effect transistor is also used in modern electronic equipment, and operation and testing of this device is covered in Chapter 8. Also covered there is a complete description of the operation of the bipolar transistor, and its uses in electronic circuits. The objective of this section is to introduce the transistor and demonstrate how simply it can be tested using only an ohmmeter. Figure 2–19 diagrams a bipolar junction transistor. Notice that it is a sandwich of two N-type materials surrounding a P-type center core. Since there are two junctions in this device, there are three leads to connect. These leads labeled E, B, and C, are called the emitter, base, and collector, respec-

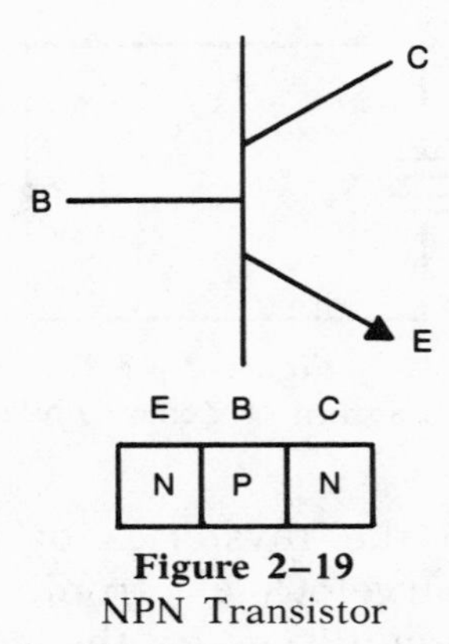

Figure 2–19
NPN Transistor

tively. Though there are many ways a transistor can be connected into a circuit, the most common hookup is as follows:

Base = signal input
Collector = signal output
Emitter = common signal ground

Figure 2–19 shows the arrangement and identifying characteristic of the NPN transistor, while Figure 2–20 diagrams the schematic of a PNP device. Note that the direction of the arrow is the identifying factor in the two devices; the NPN points out, while the PNP points in. The junctions in a transistor can be treated as if you were checking back-to-back diodes. Simply by checking the forward and reverse bias resistance of each diode junction, you can quite reliably tell the relative quality of the transistor. Figure 2–21 is a

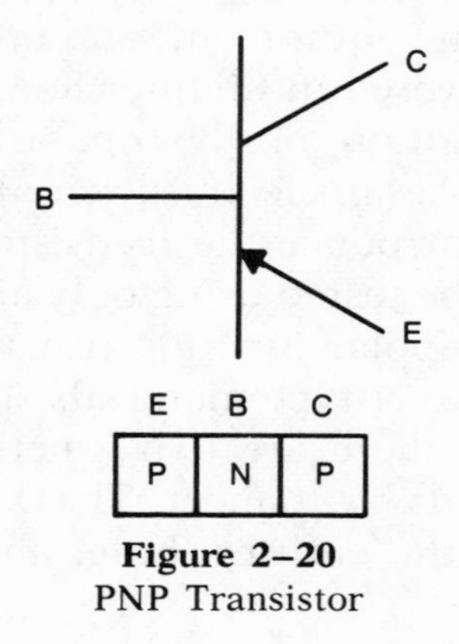

Figure 2–20
PNP Transistor

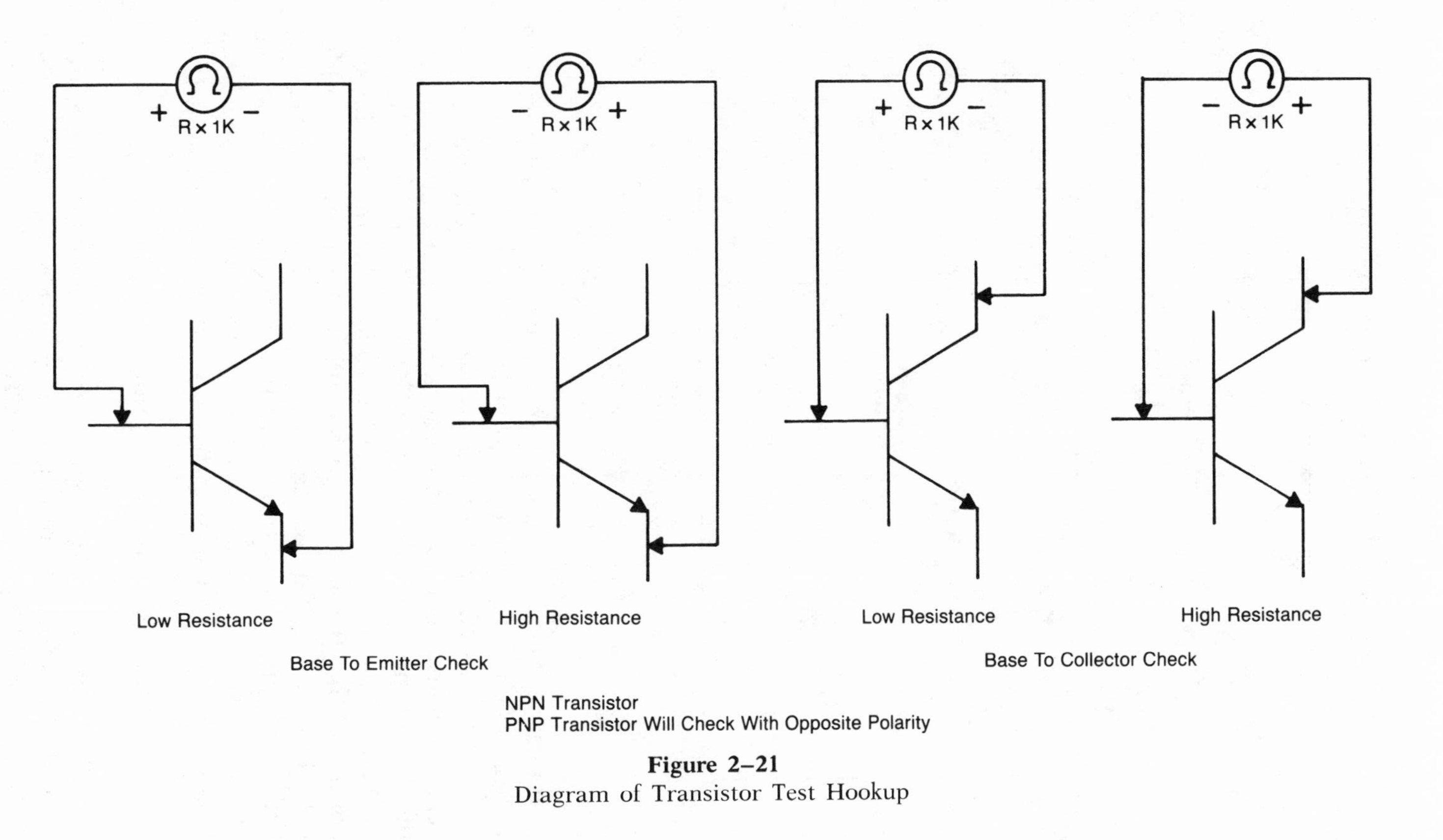

Figure 2–21
Diagram of Transistor Test Hookup

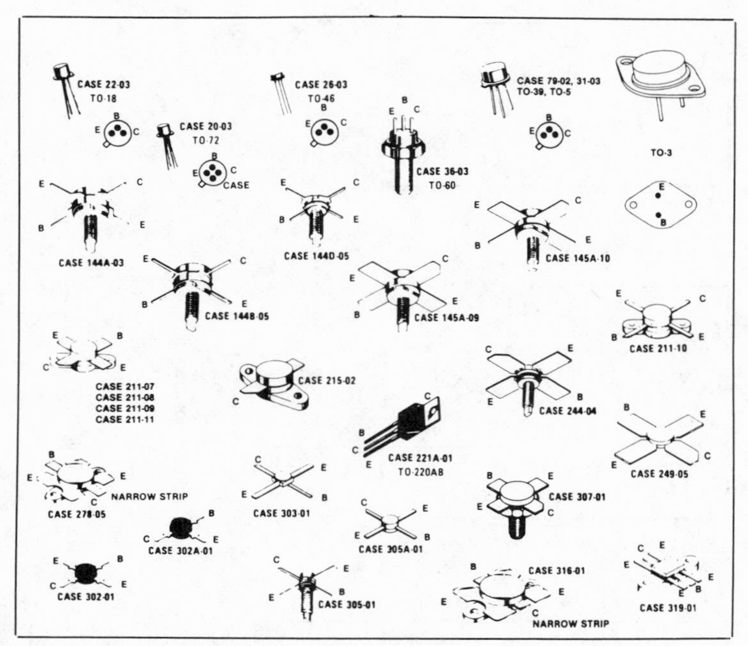

From The Radio Amateur's Handbook, *1982 edition. Published by the American Radio Relay League.*

Figure 2–22
Typical Transistor Case Styles

diagram of the proper testing procedure. The hard part is in being able to identify the three leads.

Figure 2–22 will probably be one of the most used references in this book, as it contains the case style outlines of all commonly available transistors. The base, collector, and emitter leads are clearly identified, and you only have to compare the transistor outline with the transistor under test. In some instances, case construction is identical, but lead configuration is different. Just select one diagram to check the device—if the transistor checks bad, use the other outline. If this test is negative, the transistor is defective.

What ICs Are and How to Test Them

Probably the most significant development to spin off from the space program is the development and subsequent improvement of the integrated circuit. This device has done more to reduce the size and cost of electronic equipment than any other single component. Since IC technology greatly simplifies most circuit design, most of the projects in this book are designed around these small but powerful components.

Essentially, there are two families of integrated circuits that the experimenter can work with—linear and digital. Linear ICs are designed to function in circuits such as stereo, television, communication, and other applications that require the amplification and processing of auditory and visual information. A linear integrated circuit operates like this: the output of the device closely follows the signal being put into it. For example, a linear IC amplifier, when given a small varying audio signal on its input, delivers a larger signal to its output. This larger signal is an accurate representation of the input signal with only a small amount of distortion that is natural in all amplifiers. See Figure 2–23 for a diagram of a simple IC amplifier circuit. As you can see, what goes in at the input is processed internally, and appears at the output of the device. You must supply power from a DC source for this particular device. Though there are ICs that are as simple as the one in Figure 2–23, most have other pins on them that must be connected for various functions. They may have feedback pins that are connected

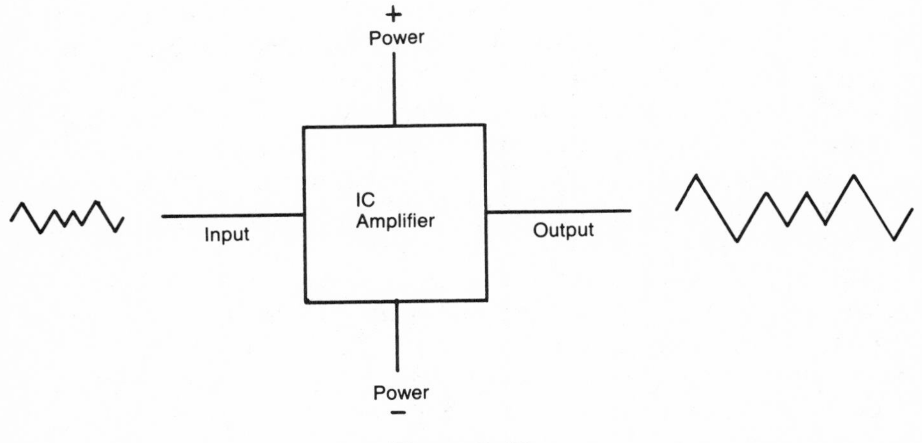

Figure 2–23
IC Amplifier

to external components to reduce distortion, or they may have control inputs to vary the gain or output voltage the device can deliver.

There are many integrated circuits that are devoted to the amplification of AC signals. These amplifiers have different functions, depending upon their application. The operational amplifier is commonly found in small signal circuits, while the differential amplifier is used whenever comparison of two signals is required. Another group of amplifiers that is relatively new is the audio power amplifier IC. This IC has become almost standard equipment in home and automotive stereo equipment, as well as in portable electronic components. The major advantage of these devices is high power output per dollar, compared to a conventional transistor circuit.

There are many specialized ICs for a myriad of functions, from generating a complex television video circuit to timers delivering accurately spaced pulses microseconds or months apart. Before we look at how these ICs are manufactured and how we can test them, let's take a look at the family of digital ICs.

The digital IC is the heart of the microcomputer. Its internal design is composed of switches, switches that can be only on or off. Compare this to the linear device that can follow an input voltage by maintaining current flow at a point between off and on at all times. The digital IC has only these two states to deal with; as a result the design can be very simple. What can be done with switches can be described in hundreds of pages, as anyone with a computer can tell you. In a digital computer, letters and numbers are represented by the on or off states of groups of eight, sixteen, or more electronic switches. These letters and numbers can be manipulated and stored in circuits called registers, or on magnetic media, such as cassette tape, or on floppy disk. An understanding of how computers operate will help you to understand the concepts of digital ICs and their purpose.

Basically, there are two types of digital circuits, called decision-making and memory. The decision-making circuits, strangely enough, are the simplest, so we will begin with them. All of these basic circuits have been constructed with transistors, diodes before them, and even vacuum tubes before them. What has made the computer, and all of the fantastic control devices and electronic games, possible is the fact that these circuits have been placed into integrated circuits.

There are five logic circuits that are the basis of digital decision making. These gates are called: AND, OR, NAND, NOR and NOT. To see how these devices do their job, let's start with the NOT gate. A NOT gate has only one input and one output. If the input has a signal that represents the ON state, the output indicates an OFF condition. This device is used for changing the state of a circuit. By the way, an ON condition is usually represented by the number one, while OFF is usually represented by a zero. Figure 2–24 is the representation of the logic gates as a digital experimenter would picture them.

Now that the NOT gate has been found an easy device to understand, let's take a look at the AND gate. An AND gate has two inputs, but only one output. Since there can be only two states at the output, a one or a zero, the AND gate decides the output in the following manner. If the input

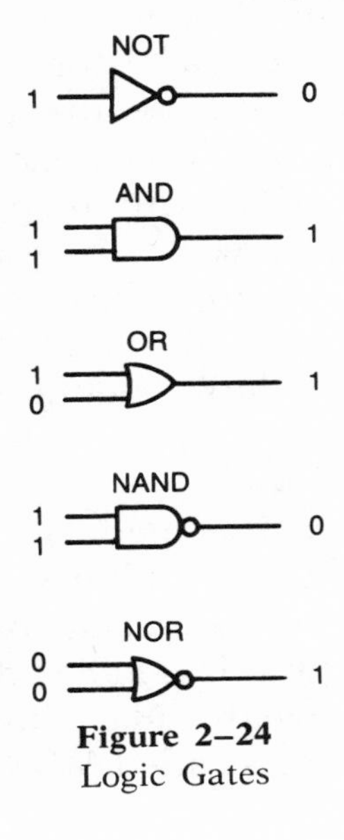

Figure 2–24
Logic Gates

states of the two inputs are both logic one, then the output will also be a one. In the case of either or both inputs being a zero, the output will be a zero. In other words, a one AND a one equal a one output. The OR gate operates in a similar manner, except that if either of the inputs is one, the output will be a one. The NAND and NOR gates reverse the condition of an AND and OR gate, respectively. In other words, they have a NOT gate at their output. If you are interested in a more detailed description of these gates, refer to a text on digital logic circuits and applications. Now, let's take a quick look at memory devices.

Memory ICs, like all digital memory media, are capable of storing either a one or a zero in each of their memory locations. An integrated circuit typically referred to as a 16K IC, will store over 16,000 individual ones or zeroes, called bits. Though 16K usually refers to 16,000 exactly, the computer industry has calculated bit storage in groups of eight. This means that there are actually 16,384 bits in a single memory IC referred to as 16K.

Let's get back to the ICs themselves. Digital circuits that are composed of single, dual, or quad gates are called SSI devices. SSI stands for small scale integration, which means that these devices contain circuitry that replaces fewer than ten or fifteen individual transistors and their related circuitry. MSI, as you might have already guessed, stands for medium scale integration. These ICs replace circuits such as registers and counters that contain up to fifty transistors. In the past, LSI, or large scale integration meant circuits containing several transistors. Today the LSI IC can contain circuitry that would need over 20,000 transistors to duplicate. These ICs are the large memory storage devices, the digital clock ICs, the microprocessor ICs, and many more specialized function devices.

The integrated circuit is made using similar technology to transistor manufacture techniques. There are bipolar and field effect types of circuitry that can be represented, though bipolar technology is practical in SSI devices only. The field effect techniques are what make MSI and LSI technology possible. This technique is called MOS technology, of which there are several types; PMOS, NMOS, and CMOS. MOS stands for metal oxide semiconductor, the N and P stand for negative and positive, while the C stands for complementary. CMOS devices contain both N and P materials.

While bipolar devices are quite hardy, and require little special care, there are handling rules that must be followed when using MOS ICs. These rules protect the device from destruction due to static electricity. Do not remove the foil or the black conductive foam from an IC until you are ready to install it. When installing a MOS IC, never touch the pins with your fingers; handle the device only by the edges. Use a socket whenever possible to avoid soldering to the component. If you must use a soldering tool on the part, be sure to use a three-wire grounded tip iron. A standard two-wire iron carries pulses from the AC line that could destroy the IC. Never subject the devices to temperatures over 120 degrees. With just these few rules, you should have no trouble handling these parts. In many years of experimenting, I have seen very few components damaged by improper handling, so a little caution goes a long way. Figure 2–25 is a photograph of several different IC case styles. The small, round device is called the metal can case, the squarish device is incased in a package called a flat pack. The long slender package with perpendicular leads is called the dual inline package, or DIP for short. This package is the most common case style in use today, and comes in the familiar black epoxy case, as well as the sturdier ceramic case. Inside these cases is a small square of silicon that contains the

Figure 2–25
Typical Integrated Circuit Case Styles

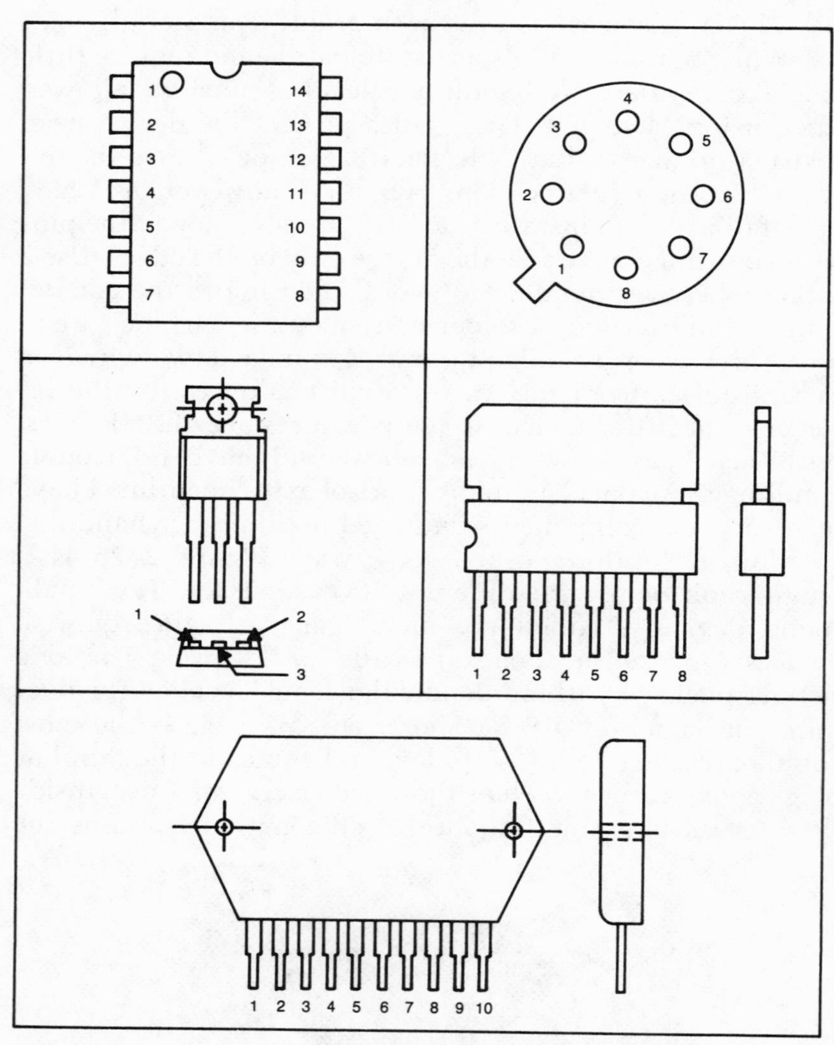

Figure 2–26
Chart of IC Cases with Pin Numbers

microcircuit. This microcircuit contains current-carrying pathways that simulate resistors, transistors, and other components. They are connected to the outside world by tiny wires that are attached to the leads of the package they are mounted in. ICs are numbered clockwise from the bottom as shown in Figure 2–26. Any IC can be tested with a voltmeter; the function of an IC must be known before it can be tested, so you must have some type of documentation on the device, such as a schematic or service manual. Another useful reference is an IC cross-reference guide, which allows you to select an adequate replacement device or determine the characteristics of the device you have. For example, if you have an amplifier IC such as that shown in Figure 2–27, you can measure each of the pins when the unit is powered up. It is important to make sure that you have power to the device, in this case on pin 8 and 4, and that you are able to verify that the other pins are all within about twenty percent of their specified value. If they are not, the IC is probably defective. If they are, you must make further checks using an oscilloscope, or the signal tracer you will complete in Chapter 11, where those techniques will be explained. We will look at all of the components discussed in this chapter in greater detail; however, let's move on to a discussion of the complete circuit, and how components are affected when connected in a circuit.

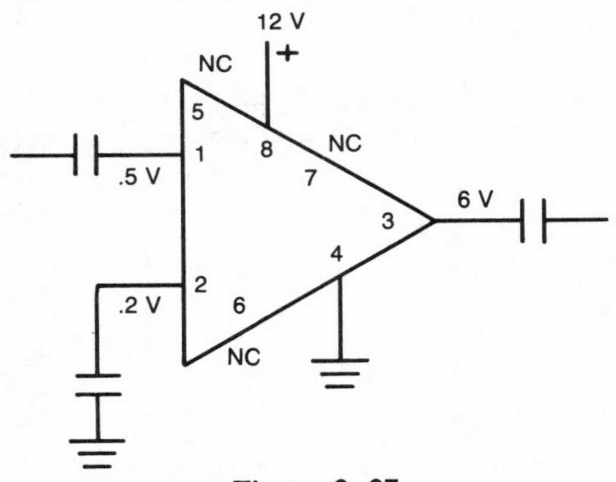

Figure 2–27
Schematic of Amplifier IC

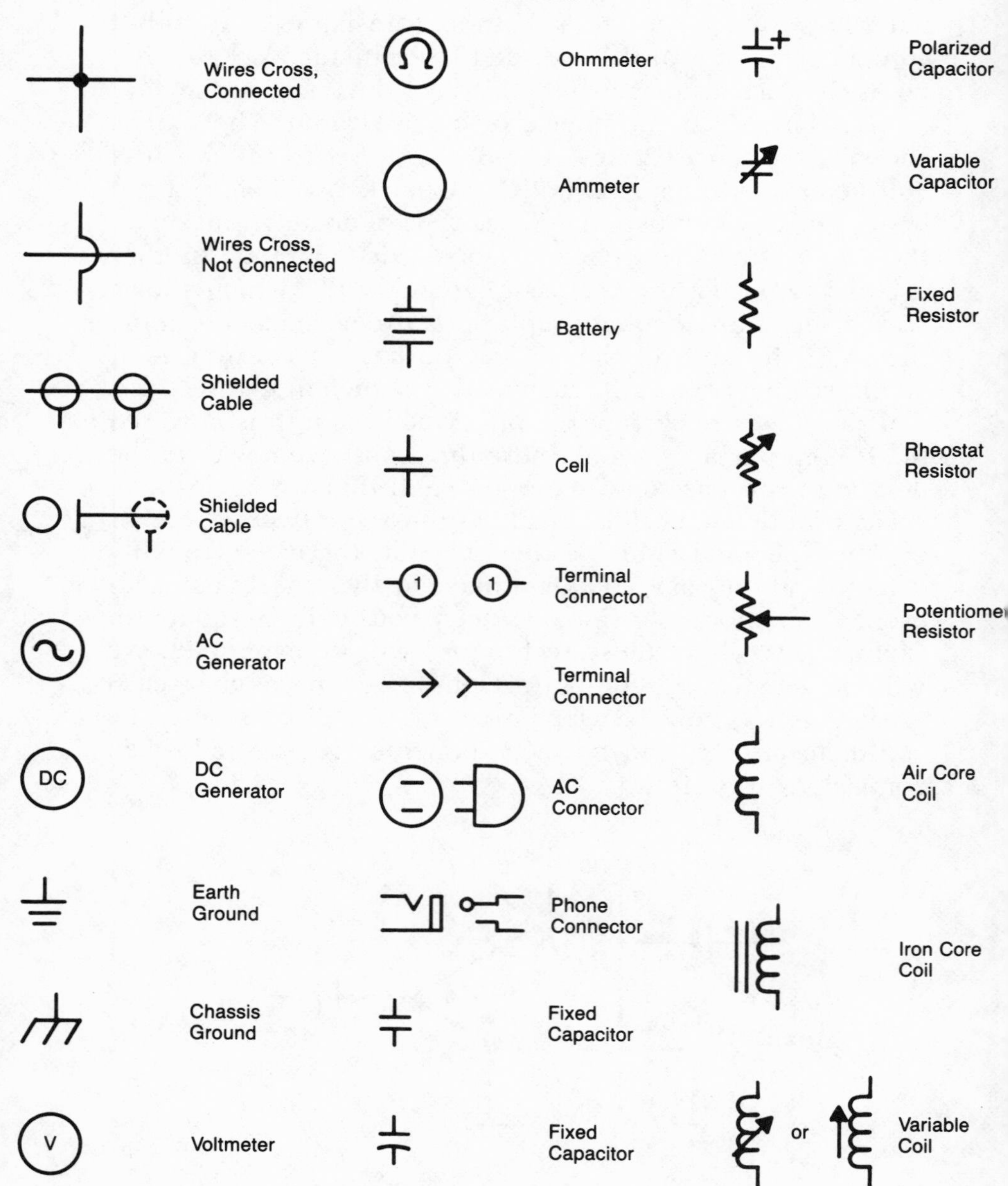

Table 3–1
Schematic Symbols

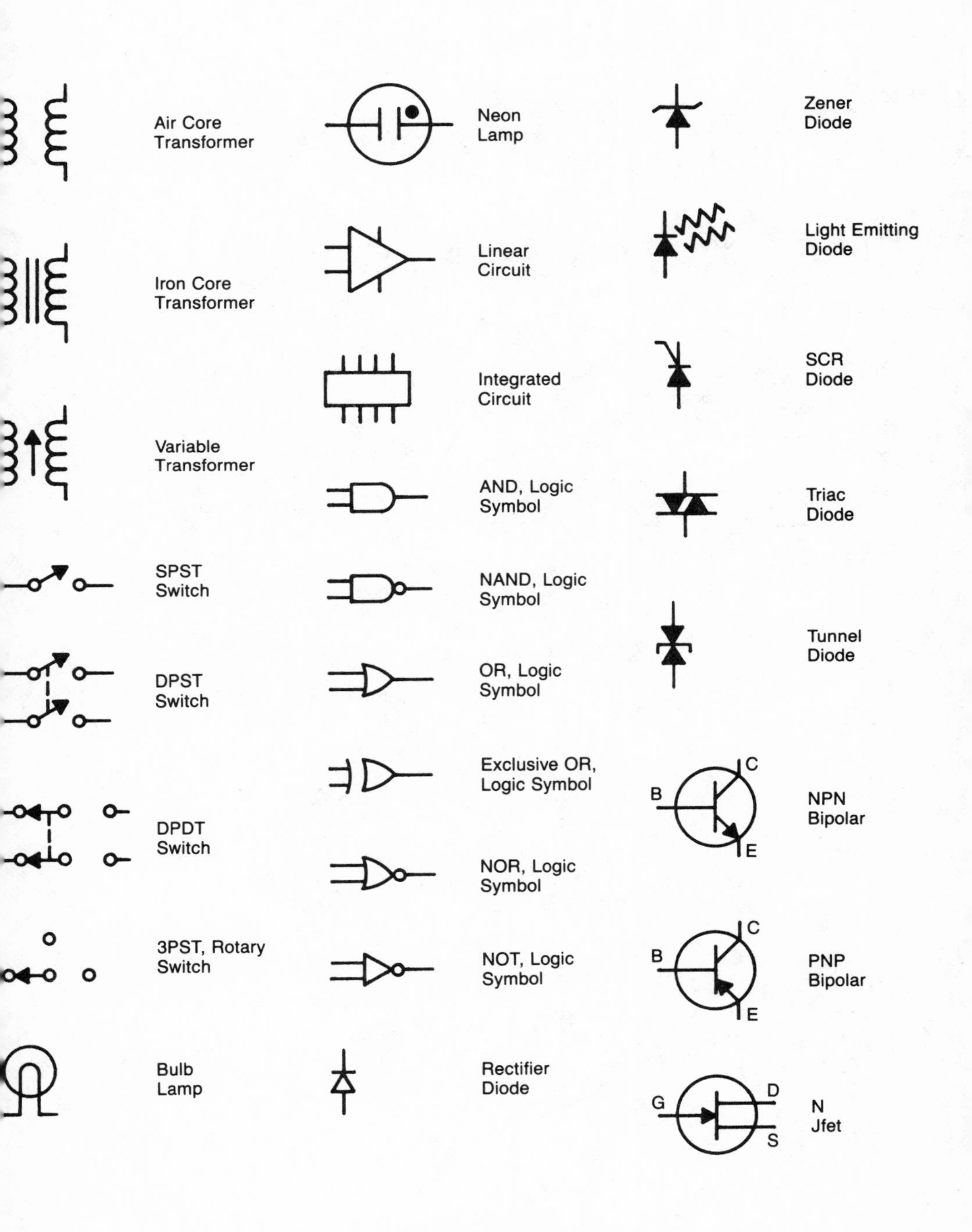

Table 3–1 (continued)

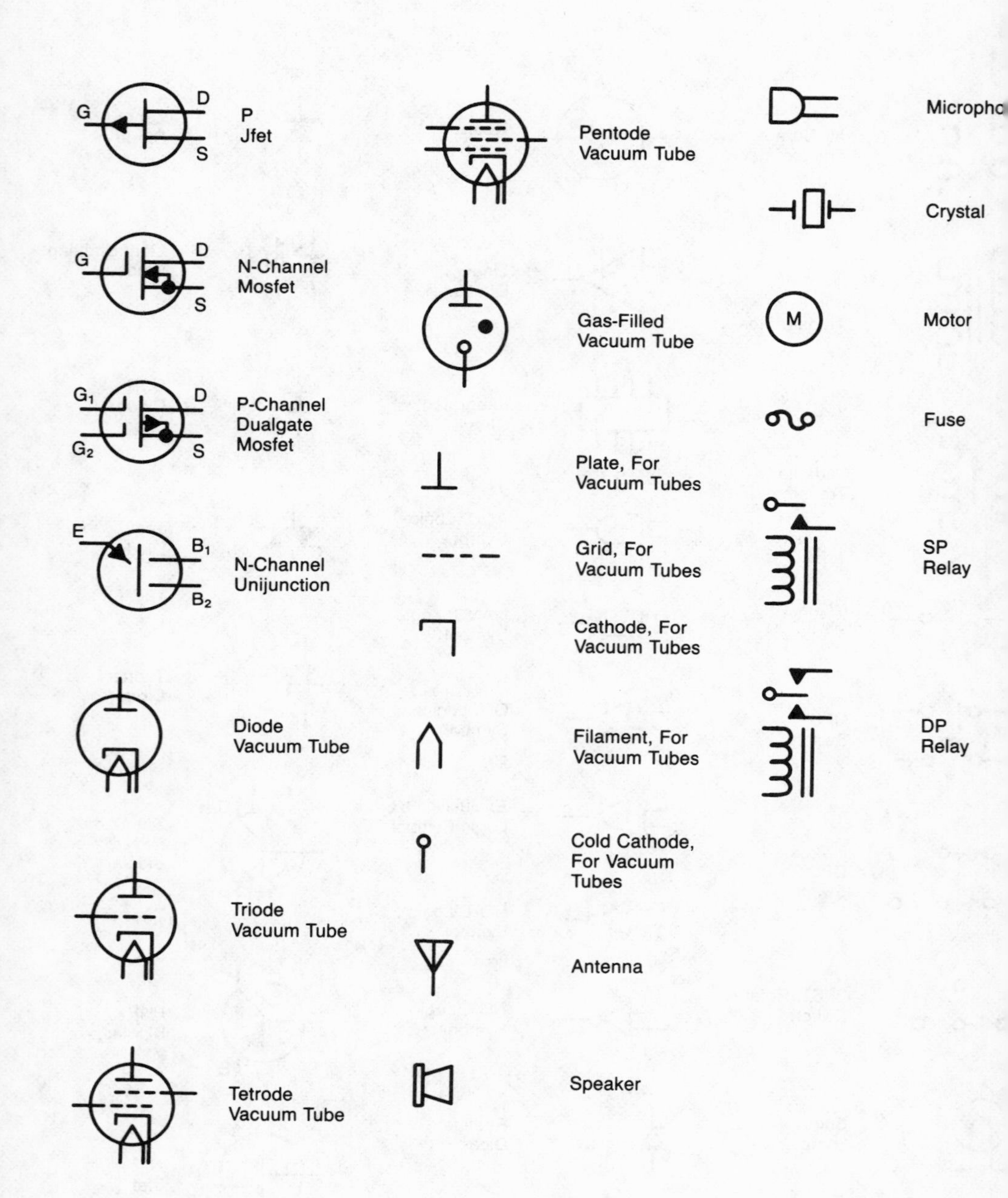

Table 3–1 (continued)

Understanding The Complete Circuit

How to Read Schematic Diagrams

You have already been introduced to some schematic symbols in the previous chapters, and the last section contained yet more symbols. Let's take a look at the purposes of schematics and their advantages. The biggest advantage of using schematic symbols is their universality. With very few exceptions, a technician can look at a schematic drawn anywhere in the world and identify the functions of the components, if not their absolute value. Another advantage of using standard symbols is that they are representatives of components, rather than pictures of components. This makes the circuits easier for the technician to draw. Look at the circuit in Figure 3–1. It is a pictorial of a circuit that includes a voltage source, a light bulb, and an on-off switch. As you can see, it is a very good representation of exactly how the circuit is to be wired. The advantages of the pictorial are ease of interpretation and the ability to construct the circuit in the exact form in which it is drawn. The beginner uses the pictorial almost exclusively because of this, but the projects are limited to those he or she has pictorials for, leaving out many projects that have only schematics shown. Figure 3–2 is a schematic diagram of the

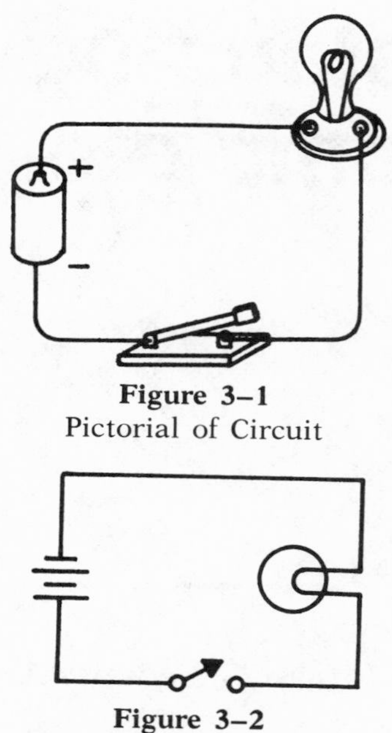

Figure 3–1
Pictorial of Circuit

Figure 3–2
Schematic of Circuit of Figure 3–1

light bulb circuit shown in Figure 3–1. Notice that the circuit is symbolized in the exact manner that the pictorial is. For example, the negative battery lead is connected to one terminal on the bulb, the other bulb terminal connects to the switch, and the other switch terminal is connected to the positive battery terminal.

The schematic representation tells the experimenter nearly as much as the pictorial does, however he or she must know just exactly what the experiments require as to layout. In most circuits, layout is up to the individual, but some circuits are critical in this respect. For example, some high frequency amplifiers require careful parts placement so that they operate correctly. This layout problem is usually mentioned by the circuit designers when they present the circuit.

There are some special circuit considerations that you should know when drawing and reading schematics. Figure 3–3 presents a circuit that shows how wires crossing other wires are represented. The dot represents a connection

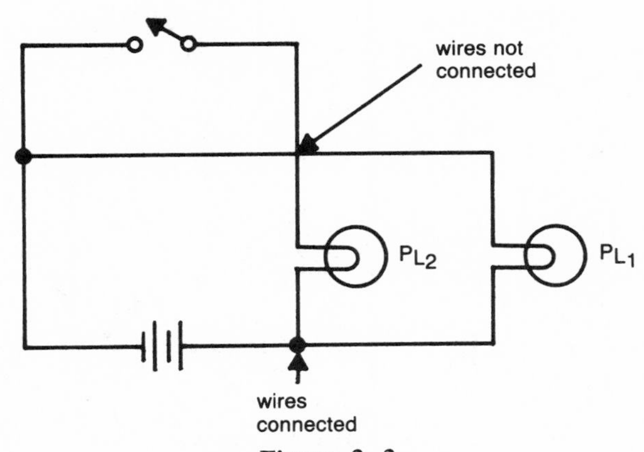

Figure 3–3
Schematic of Crossed Wires

between the two wires. The crossover above the two bulbs has no dot, therefore there is no connection between these two wires. If you trace the current flow through this schematic, you will find it follows the negative lead of the battery to a wire junction where current divides. Some current flows to the switch, while the rest travels directly to PL1. The two currents recombine at the junction of PL1 and PL2, where the current flow continues on to the positive battery terminal. Don't spend any time right now trying to understand how this circuit works, as it is only attempting to demonstrate the correct way to show proper wire connections. There is one other way wires are represented in a circuit. Figure 3–4 presents this technique. In this method, connected wires do not have a dot, while unconnected wires loop around one another. This method is usually found in older schematics, though it is still seen occasionally. If you are drawing a schematic, be sure you always use the same method. It will become second nature to you, and you won't accidentally start mixing the connection types in your schematics.

Another component that causes problems for beginners is the simple switch. Refer to the table of schematic symbols for all of this discussion. The basic switch is simple, and is referred to as a single-pole single-throw switch, which is

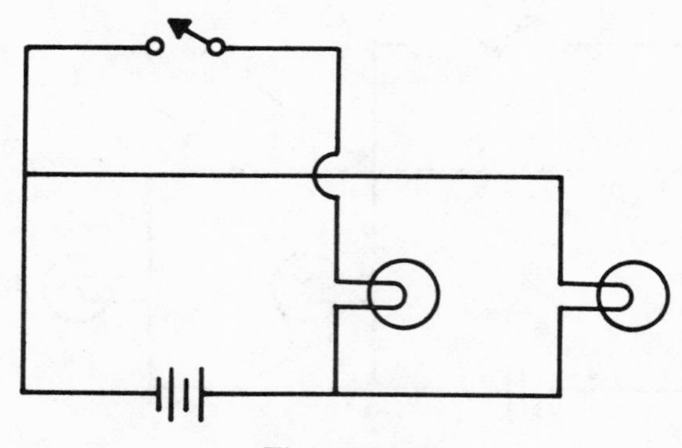

Figure 3–4
Alternate Schematic of Crossed Wires

often abbreviated as SPST. As you can see from its symbols, when it is open, the two terminals are not connected, and when it is closed, they are connected. The single-pole double-throw (SPDT) switch contains an extra pole, or contact. This simply means that the throw will either be connected to one pole or the other. The single-pole double-throw switch is actually two SPST switches ganged together, and the DPDT (double-pole double-throw) switch is really two SPDT switches connected. These switches are usually available as slide or toggle types. Rotary switches are usually multipole types; as you can see, the throw connects to several poles, one at a time. Multiple-throw rotaries are also available. These devices are several switch sections ganged to the same shaft.

Some other terms you should be aware of include the presentation of battery values, and component labeling and identification. Schematically, the battery has a short and long line, with the short line indicating the negative battery pole. Although up to now it has been represented with polarity markings, it isn't always done that way. If you find a battery with no polarity markings, just remember that the short line is negative. Figure 3–5 includes several drawings of battery schematics. The lower battery is understood to be 1.5 volts, while the center battery is 3 volts. The upper battery, though it has just as many lines, has a notice that it represents 9 volts. As you have already probably figured out,

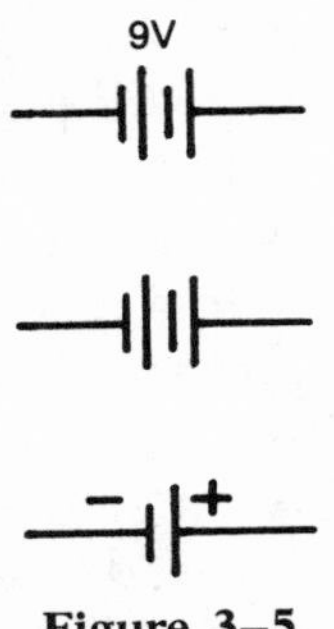

Figure 3–5
Battery Schematics

a single pair of battery lines represents one cell, or 1.5 volts. The sum of the 1.5 volt pairs then represents the total voltage; however, imagine the length of a 90-volt battery symbol. As a result, the industry has settled upon the standard of two cell symbols, with the notation as to voltage represented near the battery.

Aside from the battery, another component that has varying representations is the capacitor. If you refer to the table of schematic symbols, you will note the two styles. If the capacitor is marked + or − to indicate polarity, it is an electrolytic type, and must be connected in the circuit as shown in the schematic.

Another common representation in schematics is power supply output and ground terminals. Figure 3–6 presents a simple circuit with two ground symbols. The ground symbols indicate that the two points are actually connected in a circuit. Although in this circuit it would have been easy just to connect these two points, in more complex drawings the

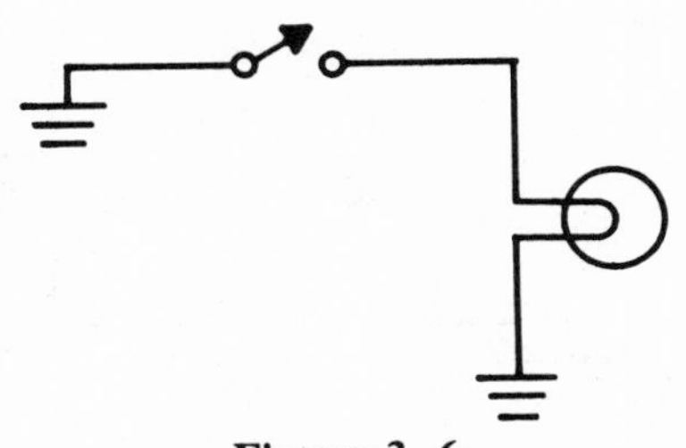

Figure 3–6
Circuit with Ground Symbols

symbol is used extensively to indicate the connection of all negative points in the circuit. All points shown connected in this fashion in the schematic are to be connected in the circuit. Figure 3–7 shows another technique when many connections are to be made to a terminal that doesn't have to be grounded. The positive battery terminal leads to a black dot (or arrow) identified with the label number 1. The bulb also connects to a similarly labeled lead. As you might guess, the two points should be wired together in a circuit. This technique is quite often used on large schematics that cover more than one page; several labels continue the circuit on the next page. Notice the ground symbol in this circuit. When you use it in this manner, you must connect that point to the common chassis point or ground point in your project. All voltages measured in this circuit will probably be referenced to this point unless otherwise directed.

Technicians usually draw their circuits starting from the left, and progressing to the right. For example, a radio schematic usually has the antenna to the left of the drawing, while the speaker is on the right. Try to follow this convention with any schematics you may draw.

Now that you have an introduction to reading schematics, let's see how to use this skill to improve our knowledge of electrical principles. The properties of voltage, current, and resistance and their effects in different kinds of circuits are covered in the next section.

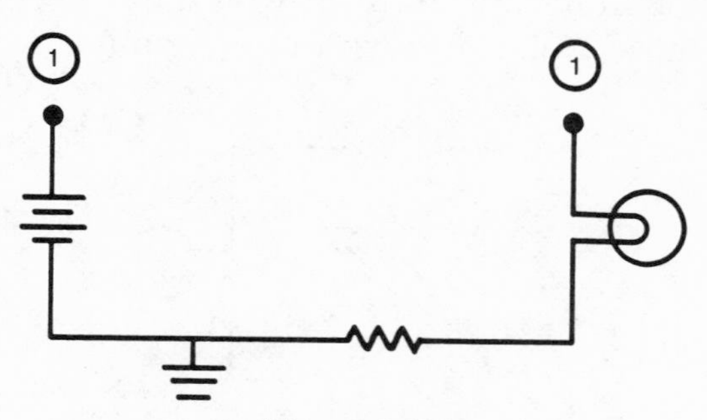

Figure 3–7
Schematic of Terminal Connections

How Voltage, Current, and Resistance Work in a Series Circuit

In Chapter 1, we experimented with a resistive circuit when learning to operate a multimeter. At that time some concepts were introduced, including the facts that voltages are additive and current is the same throughout that type of circuit. Now we will go into a little more detail about each of these concepts, as well as others that apply to this type of circuit.

Most of the circuits we have worked with up until now have been what the experimenter calls series circuits. These circuits are the simplest that the experimenter works with. Figure 3–2 is an example of this type of circuit. The series circuit has only one way for current to flow, which is always from negative to positive. Electrons travel from the minus battery terminal, through the bulb, then, after traveling through the switch, return to the battery through the positive terminal. As you can see, there is only one way for current to flow, and all components are connected end-to-end. This is the basic definition of a series circuit, any circuit that contains a three-way junction cannot be a series circuit.

Lets look at a more complex circuit. Figure 3–8 is the same circuit you built when learning to operate the meter, with the exception of the addition of a switch. If you recall from your measurement of voltages in that experiment, each resistor measured three volts, approximately, across it. Also,

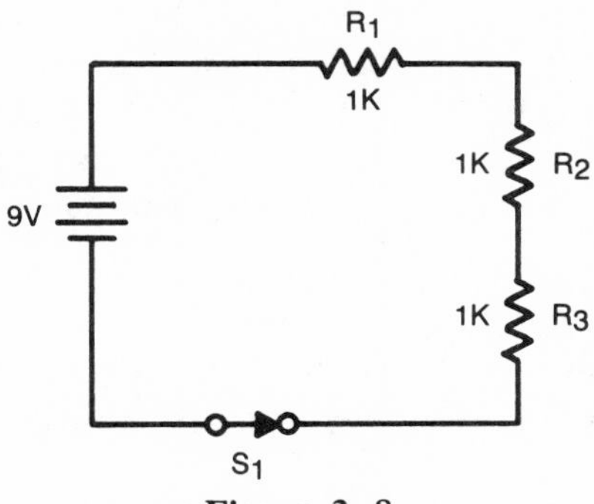

Figure 3–8
Series Circuit

the sum of these voltages should have equaled the source voltage, in this case nine volts. In other words, each resistor has a voltage drop that is in proportion to its resistance and, since all resistances are equal, so are all voltage drops. If you were to change R1 to 2 Kilohms, you would increase the voltage drop across it, while at the same time decreasing the drops across the other two. Since they are now unequal, the drops will be unequal. R2 and R3 will have lower voltages across them; however, since they are the same resistance, the two drops should match. I should point out at this time that since you are working with components that have tolerance, equals may not be exactly so. For our purposes, if two numbers are within twenty percent of expected values in a circuit, they may be considered equal. The quality of your test equipment and components will determine this, so keep in mind that all theoretical discussion of circuitry assumes that you are using perfect components. If you try to duplicate them, you will have to take this into consideration.

Until now, most of the circuits we have been talking about have been series circuits. Although they are easy to understand, they represent only a small part of the circuitry that the experimenter works with. The most typical example of a series circuit is the old Christmas tree lights that were used long ago. When one of these lights would burn out, the entire string would quit working. To find the burned-out bulb, you had to try one bulb after another until the defective one was discovered. This is probably not the best method for that type of wiring. Newer light circuits use parallel rather than series wiring. If you are familiar with the electrical wiring in your house, you are familiar with parallel circuits. Let's look at the parallel circuit and see how it changes the concepts we learned when working with series circuits.

How the Parallel Circuit Affects Voltage, Current, and Resistance

One of the most difficult concepts for the novice to grasp is the fact that many of the concepts discussed in series circuits are exactly opposite when dealing with parallel circuitry. This can be a cause for problems, so let's look first at these concepts as related to series circuits and apply them to parallel circuits. Each item in the following list will be discussed separately in greater detail.

1. All current is the same in a series circuit, while in a parallel circuit total current is the sum of individual currents.
2. The sum of all voltage drops in a series circuit equals the total voltage, while total voltage in a parallel circuit is the same.
3. Total resistance in a series circuit is the sum of the resistances, while total resistance in a parallel circuit is smaller than the smallest resistance.

Look at the circuit in Figure 3–9, which contains a simple parallel circuit. The key to understanding the operation of parallel circuits is the fact that all components are connected directly to the source. Notice that the upper end of both resistors is connected to the negative battery lead directly, while the lower ends are connected to the positive battery terminal. As you can see, both resistors have ten volts across them. Since two resistors are connected to the source, each will make its own demands upon the supply. This means that each resistor behaves as if it were connected alone, demanding one amp of current in this case. Both resistors demand one amp, so the supply must deliver two amps, and total current will be the sum of the individual currents. Two amps will flow from the negative source terminal until the current reaches the junction of R1 and R2,

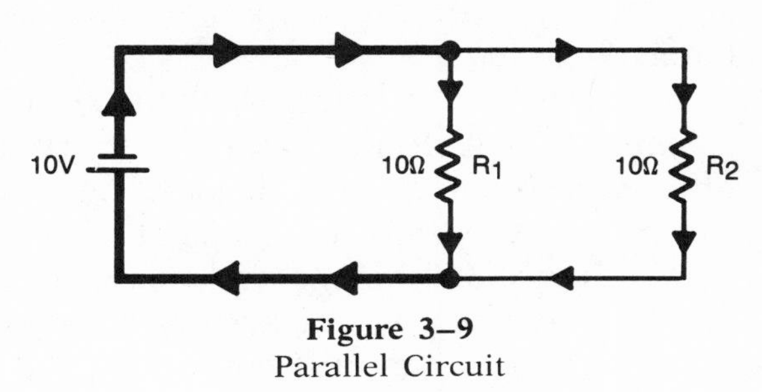

Figure 3–9
Parallel Circuit

where it divides. One amp travels through R1, while one amp continues on to travel through R2. At the lower junction, the two currents recombine and continue to the positive battery terminal. This should cover the first statement in our list above.

Earlier it was stated that both resistors were connected to the source. If this is the case, then both resistors have the total supply voltage across them, thus the second statement is also valid. Now to tackle that last statement, which said that total resistance is smaller than the smallest resistance. In our circuit, each resistor is ten ohms, and each draws one amp according to Ohm's law. We have already determined that total current is two amps. As a result, ten volts applied to a circuit that draws two amps must have a total resistance of five ohms. To prove this, hook up two resistors of the same value in parallel as in Figure 3–9, but connect an ohmmeter instead of a source. You will find the resistance measured to be one-half the value of the resistors. Changing the value of either resistor will change the total resistance; however, the total will *always* be smaller than the smallest resistor. Now let's take a closer look at circuit behavior as we rearrange, add, and remove resistors.

Keep in mind that all circuits always obey Ohm's law, so when working with parallel circuits, you may apply any rules learned earlier. In other words, current always goes up when you lower resistance, and current goes up when voltage goes up. As you can easily see, by increasing the source voltage or decreasing the value of either resistor in Figure 3–9, you will increase total current. If you change only one resistor, though, current in the other resistor will not change. Total current must change since one of the resistors changed value.

In order to assist you in learning about circuit characteristics, we can use a technique called "interrelationship analysis." This technique is a valuable learning tool that will assist you in understanding the concepts of series and parallel circuits. Figure 3–10 is a simple series circuit, with a series of boxes drawn below. The arrows drawn inside the boxes represent a change in circuit operation. The up arrow

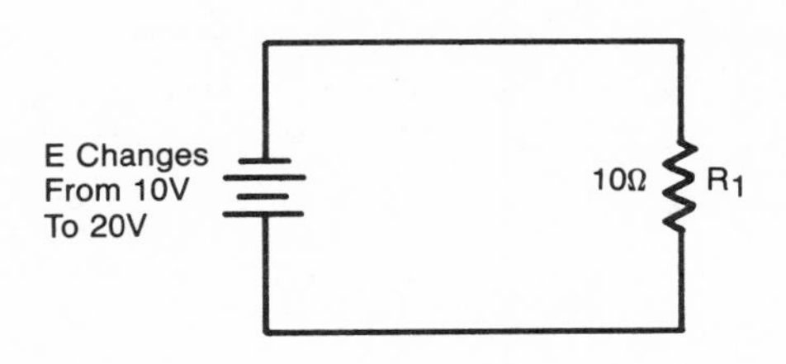

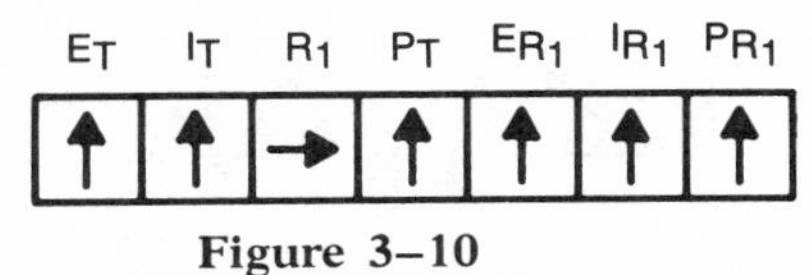

Figure 3–10
Interrelationship Analysis

↑ indicates an increase in a circuit parameter, the down arrow ↓ obviously designates a decrease in that value, and the right arrow → indicates no change. In Figure 3–10, the left box indicates an increase in Et, total voltage. This technique always uses the left box to indicate the varied item in the relationship. Always vary only one item at a time when using this technique; otherwise it is easy to become confused as to proper circuit operation. As we already know, if voltage increases, current does also, therefore the box labeled "It" also has an up arrow. Since the actual resistor was not changed, the box labeled "R1" has a right arrow. Remember, resistors do not change value unless you change them. Since power is the product of voltage times current, Pt must also have increased. The last three squares in this circuit apply to the particular component parameters of R1. Even though we did not change the resistor value, all circuit parameters that apply to the resistor did change.

Let's analyze a series and parallel circuit, then you can try a few on your own. Figure 3–11 is a series circuit containing two resistors and a battery. As you can see, we will analyze what happens when R1 is replaced with a larger value resistor. To make the exercise more concrete, let's say we replaced R1, a 1000-ohm resistor, with a 2000-ohm resistor. R2 is 1000 ohms and Et is a 9-volt battery. If you have trouble with the analysis, just wire the circuit and

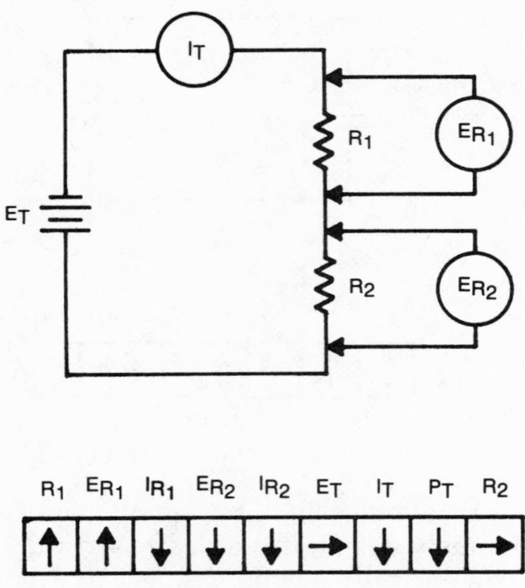

Figure 3–11
Series Circuit

measure its values, or use Ohm's law (Chapter 4) to calculate them. As you already know, when you increase the value of a resistor, you increase the voltage drop across it, therefore Er1 goes up. Also, since total resistance increased (remember, Rt = R1 + R2), current must have gone down. Now comes the hard part—we did not change R2, yet the Er2 arrow points down, indicating a decrease in voltage. How can this be? If you recall, the source voltage is nine volts since we did not change this voltage, and since the drop across R1 increased, the drop across R2 must have decreased. Unless we supply extra voltage to keep Er2 constant, it must decrease. Since total resistance went up, and current in a series circuit is the same, then Ir2 must have decreased. Et remains constant since we did not change the battery voltage, and power decreased since current went down. R2 remains constant, since we did not change it. Now let's examine a parallel circuit.

In Figure 3–12, we are going to lower the value of R1 from 2000 ohms to 1000 ohms. R2 is 2000 ohms, and the

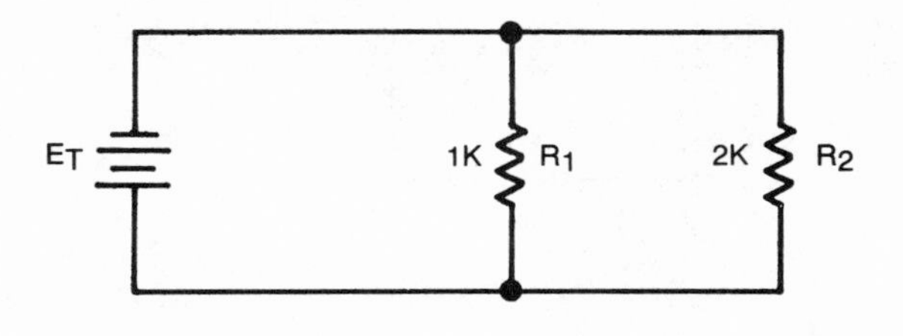

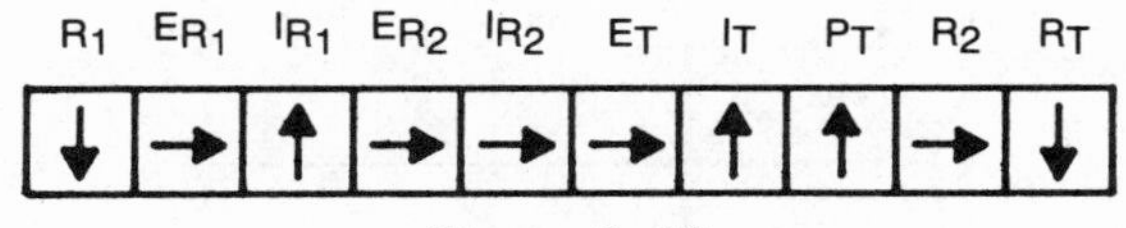

Figure 3–12
Parallel Circuit

battery voltage is 9 volts. This is a parallel circuit; therefore the total supply voltage is across both resistors, and as a result the voltage across R1 (Er1) won't change. Branch current will increase, since R1 decreased in value. Er2 remains constant, as do Ir2 and Et. Since branch current increased in the first branch and It equals Ir1 + Ir2 in a parallel circuit, total current will increase, as will total power. Since R2 was not changed, its value remains constant; however Rt decreases, since the total resistance in a parallel circuit is always smaller than the lowest resistor in use. Now that you see how it is done, Figure 3–13 contains three practice circuits to analyze. Cover Figure 3–14 with a sheet of blank paper, then lower the edge of the paper until you can see the labels above the boxes. Draw the arrows on the edge of the sheet of paper, then check your answers with those in the figure by lowering the paper a bit more and comparing your answers with ours. If you have trouble understanding the concepts presented, review the first part of this chapter, or better yet, wire and measure the circuits using actual components. There is no substitute for practical experience. When you feel confident that you can understand both series and parallel circuits, go on to the combination circuits in the next section. These circuits contain both series and parallel elements, and are representative of most practical circuits that you will actually use. (See p. 98.)

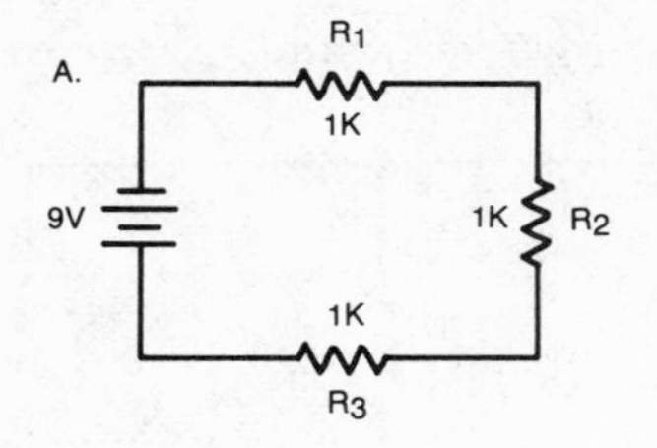

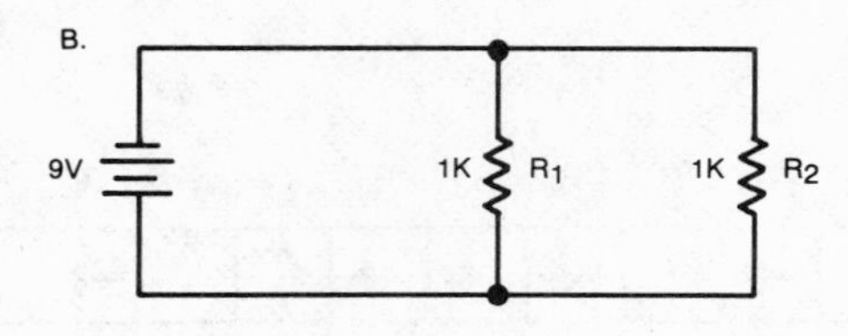

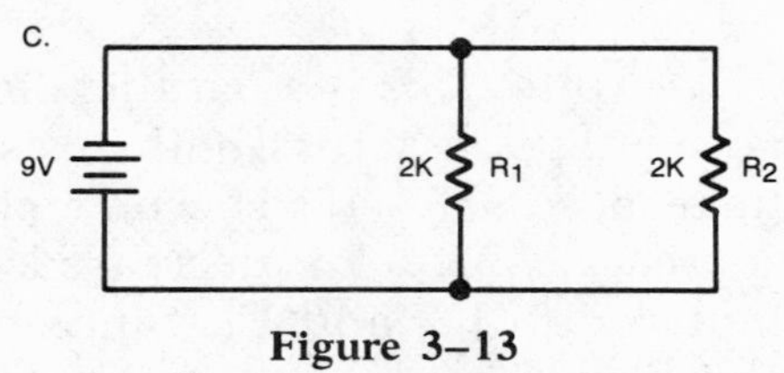

Figure 3–13
Three Practice Circuits

A. R_1 is replaced with a 2K resistor.

R_1 R_2 R_3 E_{R1} E_{R2} E_{R3} E_T I_T R_T P_T

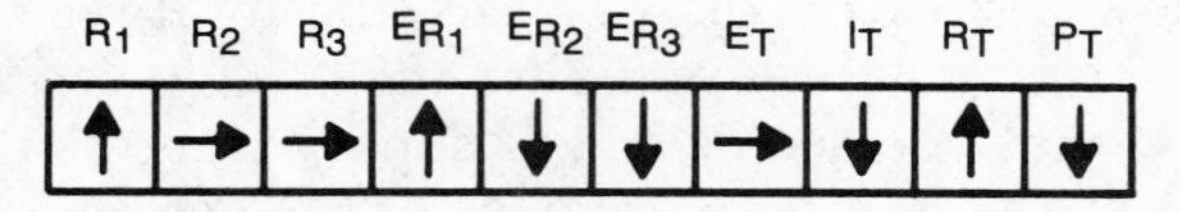

B. R_1 is replaced with a 4K resistor.

R_1 E_{R1} R_2 E_{R2} E_T I_{R1} I_{R2} P_{R1} I_T P_T

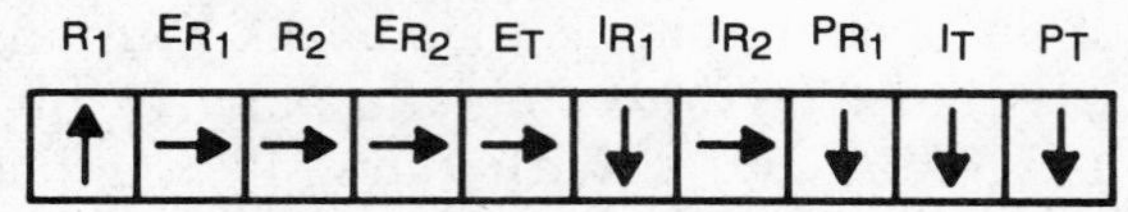

C. E_T is increased to 18 volts.

E_T I_T R_T E_{R1} E_{R2} I_{R1} I_{R2} P_{R1} P_{R2}

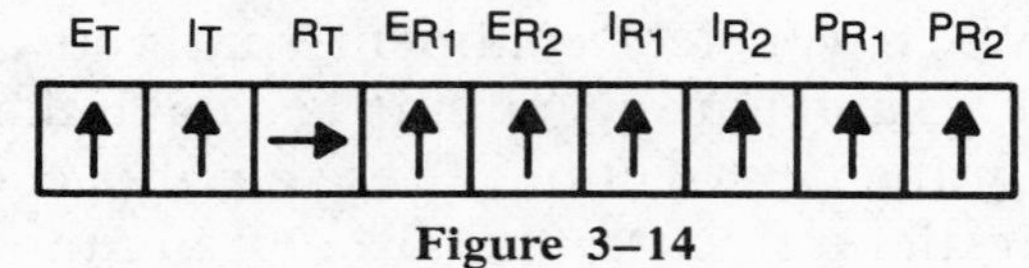

Figure 3–14
Solutions to Practice Circuits

How the Combination Circuit Affects Voltage, Current, and Resistance

Practical circuits usually contain many series paths and many parallel paths. As a result, they become more difficult to analyze. If you keep the following rules in mind, however, you should be able to keep them straight. **Rule 1:** In a combination circuit, treat all series elements as a series circuit. **Rule 2:** In a combination circuit, treat all parallel elements as a parallel circuit. With these two rules in mind, let's look at a few examples. Figure 3–15 contains a simple combination circuit. All resistors are 1000 ohms, and the battery is 9 volts. When analyzing circuits, always start from the negative battery lead. The electrons first travel through R1. Since all electrons must pass through this resistor, this must be a series component, and current through R1 will equal total current. Immediately after leaving R1, the current divides between R2 and R3. As these are equal, the currents will be equal in each branch. As with all parallel circuits, Ir2 + Ir3 equals It. The electrons recombine and continue on to the positive battery lead. Total current (and current through R1) can be measured at the points marked A, B, and C. R2 and R3 branch currents can be measured at points D and E, respectively.

Total resistance can be calculated easily in this circuit. The equivalent resistance of R2 and R3 will be exactly one-half their rated value, or 500 ohms. This Req (equivalent resistance) is in series with R1, therefore total resistance in this circuit is 1500 ohms.

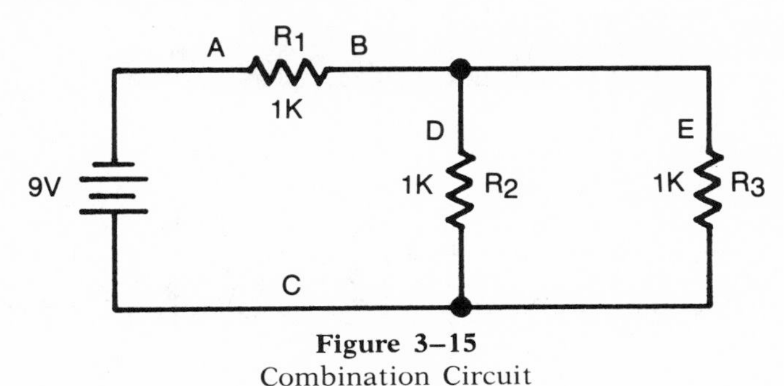

Figure 3–15
Combination Circuit

Now let's take a look at voltage drops in the combination circuit. There are two drops in this circuit, the drop across R1, and the drop across the branches R2 and R3. Remember, the voltage drop is the same in a parallel circuit, while it is additive in a series circuit. Since the Req of R2 and R3 is 500 ohms, and R1 is 1000 ohms, the voltage drop is proportional to the total resistances. In other words, 6 volts will appear across R1, and 3 volts will appear across the parallel branch.

Let's look at another circuit. Figure 3–16 is a combination circuit containing two parallel branches and a series resistor. This circuit is easily analyzed by converting each parallel resistance to its equivalent value and analyzing it as a series circuit. As all resistances are 1000 ohms, the total resistance of each parallel branch is 500 ohms. The total

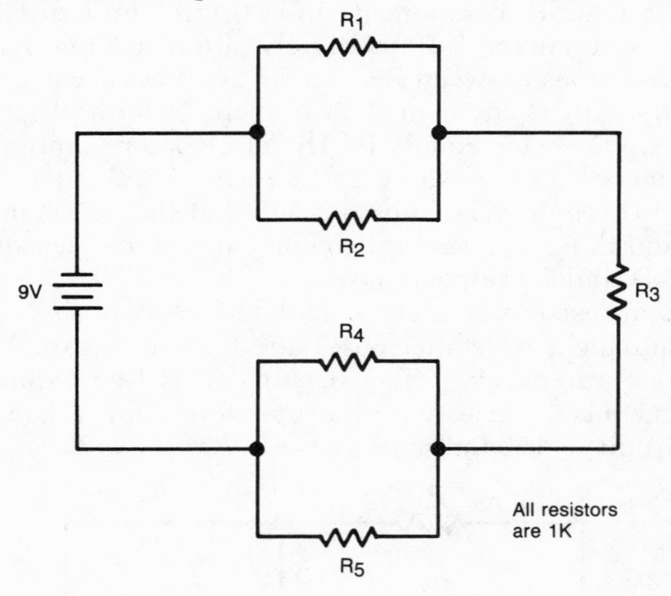

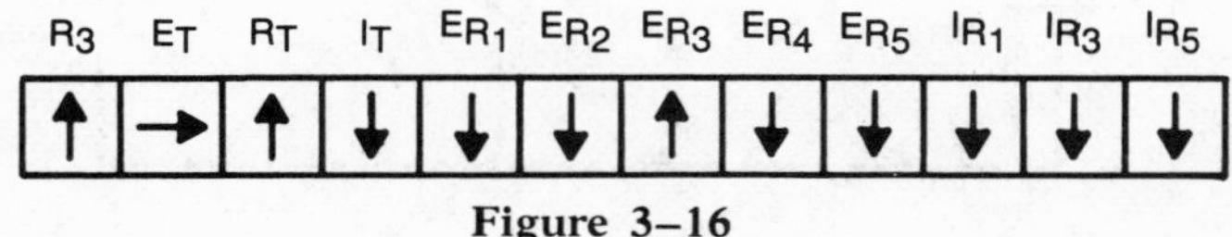

Figure 3–16
Combination Circuit

circuit resistance is then 2000 ohms. (The Req of R1 and R2 + R3 + the Req of R4 and R5.) With this in mind, let's do an interrelationship analysis. Assume R3 is increased to 2000 ohms. Et will remain constant, and Rt will increase. Because of the increased circuit resistance, It will decrease. The voltage drops across all other resistors must decrease because the voltage across R3 will increase. Since total current decreases, all currents in the circuit also decrease.

Look at Figure 3–17, which has a series circuit inside a parallel branch. The first action taken by electrons leaving the negative battery terminal is dividing at the junction of R1 and R2. Total resistance in the branch containing the series resistances is 2000 ohms, just as is the first branch. Current then divides equally and recombines at R4. The parallel branch total is 1000 ohms, therefore total resistance is 2000 ohms (Req + R4).

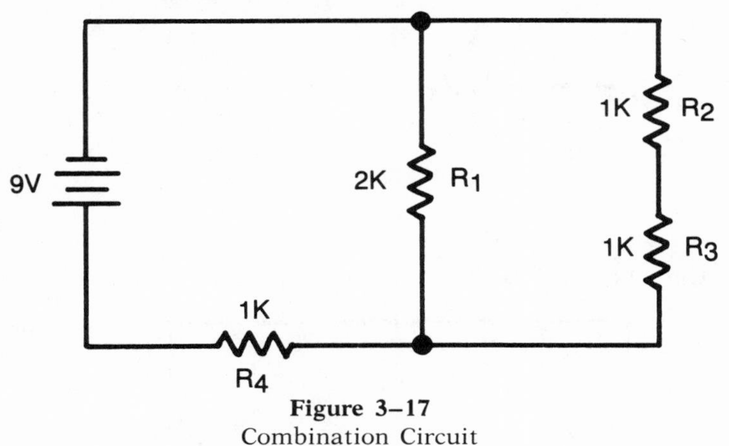

Figure 3–17
Combination Circuit

Figures 3–18 and 3–19 contain some more interrelationship analysis circuits and their solutions, respectively. Try to analyze these problems by breaking them down into their series and parallel components. If you cannot understand exactly what is happening in these circuits, build them and take some measurements to see what happens. If you still have problems, review this chapter, and read Chapter 4. Chapter 4 will give you further insight into calculating and working with circuits mathematically.

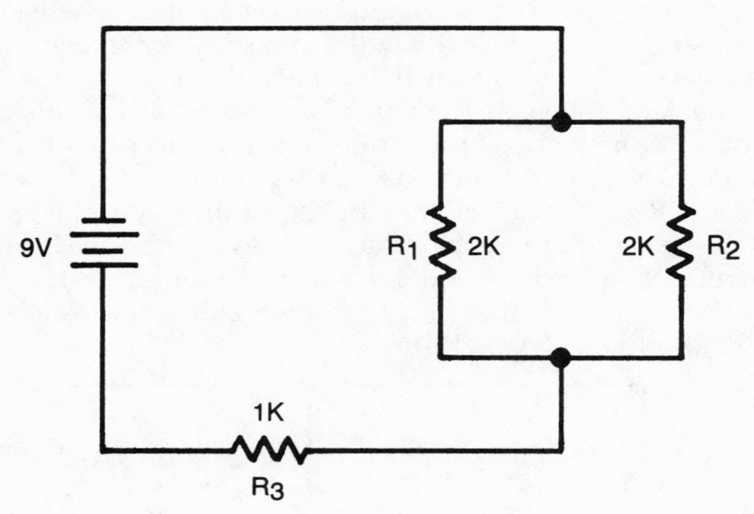

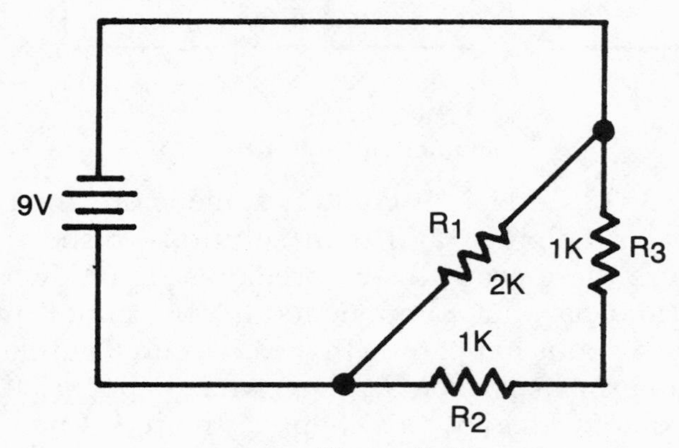

Figure 3–18
Interrelationship Analysis

A. R_3 is increased to 2K.

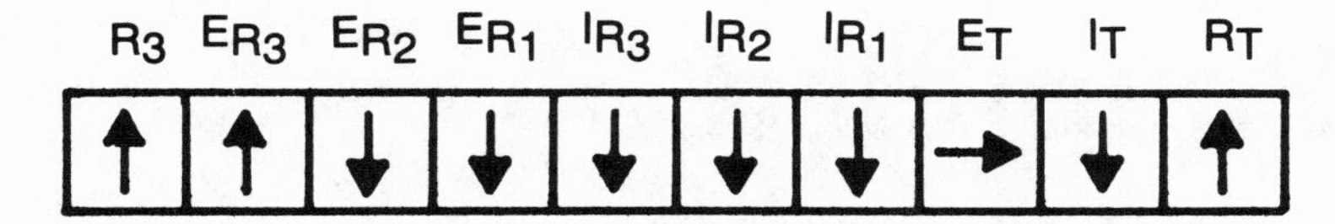

B. R_1 is decreased to 1K.

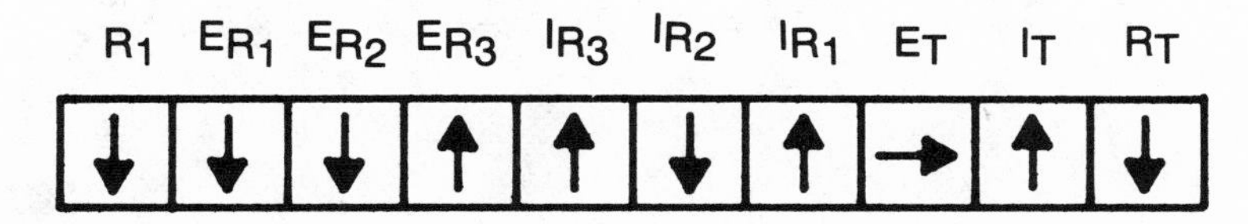

C. E_T is increased to 18 volts.

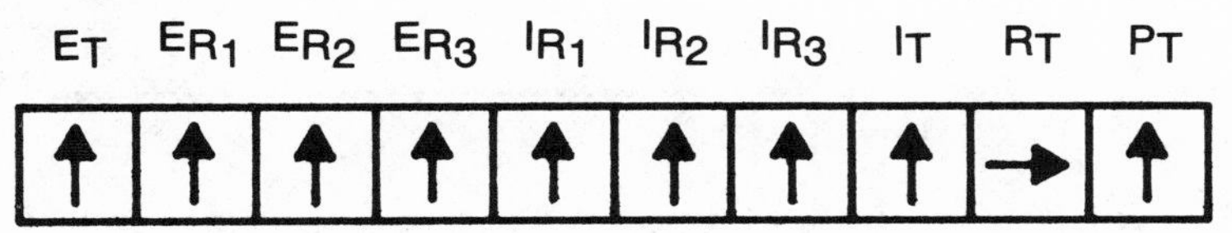

D. E_T is decreased to 6 volts.

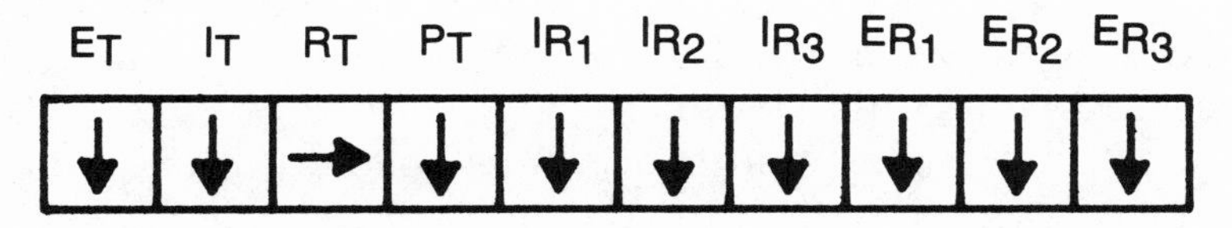

E. R_1 is increased to 4K.

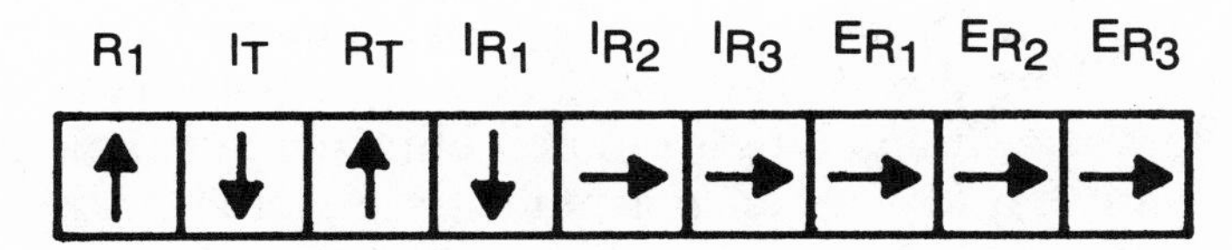

F. R_3 is increased to 2K.

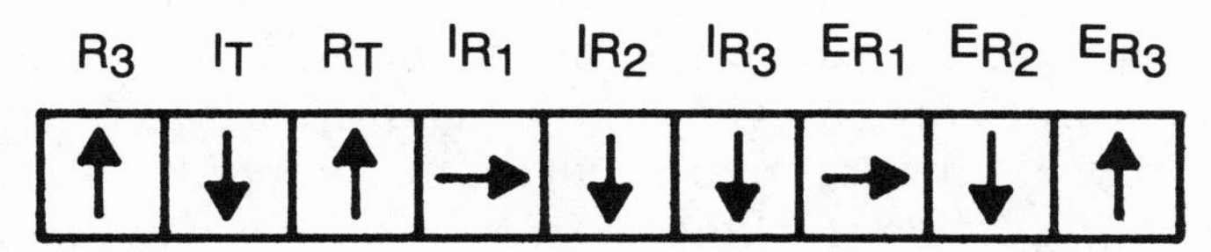

Figure 3–19
Solutions to Above Circuits

Simplified Approaches To Electronics Math

4

Working with electronic circuits is interesting; however many people get bogged down when they have to start working with electronics math. The procedures illustrated in this chapter should assist you in overcoming this hurdle. If you already feel comfortable with electronics math, feel free to bypass this chapter.

All laws, rules, and procedures covered in this chapter can be learned with only the assistance of a pocket calculator. The calculator has become, in a few short years, one of the most valuable tools of the experimenter. For many years, electronics books and courses spent many pages and hours discussing the uses and functions of the slide rule. The slide rule was seen hanging from belt clips of many students in electronics, and was an extremely useful tool. The calculator now performs more functions with far greater accuracy than the slide rule, and can now be found hanging from belt loops, replacing the slide rules of a few years ago. If you do not have one, don't go out and spend a lot of money for a scientific type, as you won't need all of its features.

One useful function that a calculator should have, if you plan on using it for working electronics circuits, is square root. The square root function is required in many formulas, and solving for it is quite difficult; you need to have a calculator with that function. Treat the calculator as a tool of the experimenter, don't feel inadequate if you cannot solve problems without one. You shouldn't feel any worse about using a calculator than you would about using needle-nose pliers, or any other tool. You will find it one of the most useful tools you have.

The first step for the experimenter in working with electronics math is to remember one basic rule. You must convert all metric prefixes to their basic unit value. This means that all units beginning in milli, micro, kilo, or mega must be converted. Table 4–1 lists the formulas required. One example should be sufficient to explain the use of the

Prefixes of Standard Metric Units

Name	Symbol	Amount	Power of 10
exa	E	1 000 000 000 000 000 000	18
peta	P	1 000 000 000 000 000	15
tera	T	1 000 000 000 000	12
giga	G	1 000 000 000	9
mega	M	1 000 000	6
kilo	k	1 000	3
milli	m	0.001	−3
micro	u	0.000 001	−6
nano	n	0.000 000 001	−9
pico	p	0.000 000 000 001	−12
femto	f	0.000 000 000 000 001	−15
atto	a	0.000 000 000 000 000 001	−18

Note: Prefixes between kilo and milli have been omitted because they are not used in electronics.

Table 4–1
Metric Equivalents

chart. You must convert 1 Kohm to its basic unit. The chart states K (kilo) = units (in this case ohms) multiplied by 1000. Our 1 Kohm resistor is thus 1000 ohms.

Ohm's Law Simplified

In earlier chapters, we looked at Ohm's Law and discussed some of its applications. Now let's take a closer look at this rule, with an eye toward understanding exactly how an experimenter can put these rules to use. In its simplest form, Ohm's Law states: One amp of current will flow when you apply one volt to a one-ohm resistance. As you can see, there is a relationship here that can be calculated if we know any two of the three parameters. Figure 4–1 illustrates what I call the Ohm's Law wheel. The E in the upper half of the wheel represents voltage, as usual.

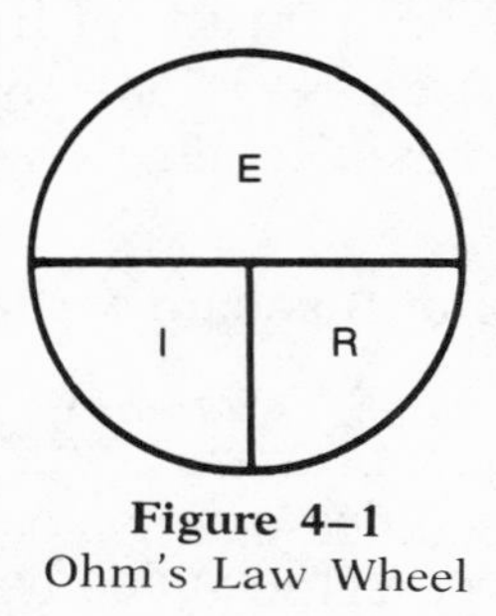

Figure 4–1
Ohm's Law Wheel

The I and R represent current and resistance, respectively. This wheel can be used at any time to solve for any given characteristic of a circuit.

To use the wheel, let's study some practical examples. If you want to use a LED (light emitting diode) as a pilot indicator for a project, you cannot exceed the maximum current of the diode. This current is specified in the data sheet for the device, though there are very few LEDs that would be harmed if you applied 20 ma. You cannot just hook the device to the twelve volts that your power supply puts out, or it will probably burn out. The correct solution to this problem is to install a series resistance that will limit the current to a safe value. Figure 4–2 demonstrates what this circuit will look like. Ohm's law will allow us to determine the value of resistance if we know the voltage and the current. You know the source voltage is 12 volts and the other voltage you must know is the voltage that will appear across the diode. You must get this from the manufacturer to be sure; however, most general-purpose LEDs are 1.2 volts. We will use this value in our example. Ohm's law states that R (resistance) is equal to E (voltage) divided by I (current). To use the wheel, just cover the desired unit with your hand.

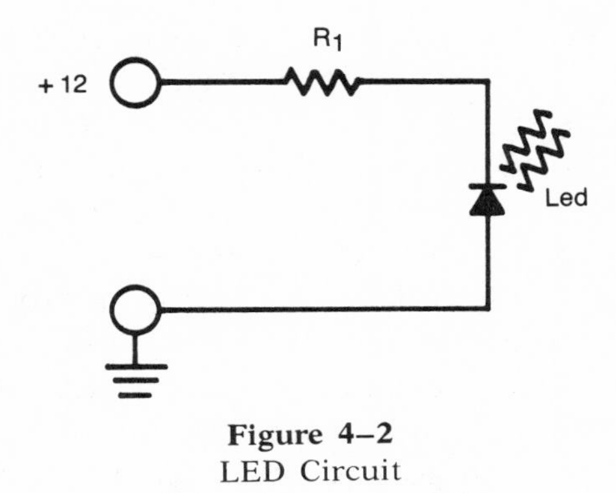

Figure 4–2
LED Circuit

The circle plainly shows the correct formula: R equal E over (divided by) I. For our example, the voltage across the resistor is 12 minus 1.2, or 10.8 volts; and the current we desire is 20 ma., or 0.02 amps. Remember to convert all units smaller or larger than the basic unit into this figure. In this problem, 20 ma. must be converted to 0.02 amps. The correct solution to this example is as follows:

R1 = Er1/Ir1

Therefore R1 = 10.8/0.02, and entering these numbers into your calculator you will find that R1 should be 540 ohms. As with all electronic circuits, when you calculate a value, it is nominal. This means that you will probably not be able to purchase a 540-ohm resistor; however, if you buy the next closest standard value, 560 ohms, the circuit will work. It will draw not quite 20 ma. of current any more, since the resistance is higher than calculated. To see exactly what the current flow would be, let's turn the formula around and solve for I. Covering the I in the wheel, the formula shown is

I=E over (divided by) R.

Substituting the values we have I = 10.8/560, which equals 0.0192 amps, or 19 ma. This should certainly be close enough to our desired current, and the circuit will work as designed.

Let's try another example. You have found a pilot lamp in an old circuit, and would like to use it in a project, but don't know its voltage. The bulb is stamped with its current rating of 50 ma. To solve for voltage, you must have the other two items, so you measure the resistance of the bulb with your ohmmeter. It measures 120 ohms, so you can now calculate the bulb voltage. Cover the E in the wheel, and you can see that E is equal to I multiplied by R. When two units are close together, with no multiplication sign, such as IR, it is understood that the function to be performed is multiplication. In this case, 0.05 amps times 120 is equal to 6 volts, so we must have a 6-volt bulb. I should point out that it is difficult to calculate the values for a pilot light. This is because the filament does not have a constant resistance. As the bulb changes temperature, its individual resistance changes, thus changing the calculated value. Use this exercise only as an example.

One last example in using the wheel; you want to build a circuit that will run on batteries. By using Ohm's law, you can determine what size batteries would be best for your application. If you choose too large a battery, the weight and size of the project will be unnecessarily large, yet inadequate batteries will require frequent replacement. The circuit is designed to operate at 6 volts, and its resistance is 1200 ohms. Let's calculate the current this circuit will draw. Covering the I, we find that

$$I = E/R, \text{ or } I = 6/1200$$

Entering those numbers into the calculator, we find the current to be 0.005, or 5 ma. Consulting a reference source on batteries, we find that a battery of four AA cells will provide the current capacity we need.

Another law—actually part of Ohm's law—is Watt's law. This law is used to find exactly how much power a circuit requires and the power wheel is pictured in Figure 4–3. Its use is the same as that of the other wheel. P (power)

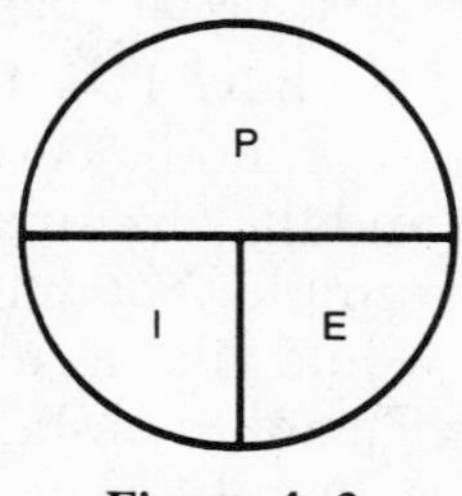

Figure 4–3
Power Wheel

is equal to I (current) multiplied by E (voltage). You can solve for any unknown if you know the value of the other two. Just one quick example—how much current does a 100-watt light bulb draw? Covering the I, we find

$$I = P/E, \text{ or } I = 100 \text{ watts divided by } 120 \text{ volts.}$$

Plugging those numbers into the calculator, we find that

$$I = 0.83 \text{ amps or } 830 \text{ ma.}$$

There is a problem with the power formula. We cannot directly calculate power if we know only current and resistance, or voltage and resistance. One solution to this is to use the Ohm's law wheel for the first part of the calculation, and the power wheel for the final step. This is really unnecessarily complicated, though, if you can remember the following two formulas:

$$P = I^2 \times R \text{ and } P = E^2/R.$$

Examples of these problems will be found throughout this chapter.

Up until now, we have looked at these laws and how they relate in complete circuits. In other words, we have been dealing with total voltage, current, and resistance in a circuit. The next three sections will detail how to solve for individual values in circuits. Learning this skill will be quite important in understanding how individual components function in a given circuit. Table 4–2 is included here to

Commonly Used DC Formulas

Ohm's Law	Watt's Law
$E = I \times R$	$P = I \times E$
$I = E/R$	$I = P/E$
$R = E/I$	$E = P/I$
	$P = I^2 \times R$
	$P = E^2/R$

Series Circuits	Parallel Circuits
Et = Er1 + Er2 + Ern	Et = Er1 = Er2 = Ern
It = Ir1 = Ir2 = Irn	It = Ir1 + Ir2 + Irn
Rt = R1 + R2 + Rn	Rt =

Table 4–2
Commonly Used Formulas

provide you with a quick reference to the most commonly used formulas. Now let's take a look at problem solving in series circuits.

Solving for Voltage, Current, and Resistance in Series Circuits

You already know how to solve for total voltage in any circuit if you know the total current and resistance. By the use of the Ohm's law wheel you can determine other factors as well. As we talk about each circuit, I will list the known parameters, and you can see exactly how these circuits can be analyzed as to operation. The circuit in Figure 4–4 is a simple series circuit. There are no values specified in the schematic except for the part reference numbers. By drawing the circuit without values, we can use the same series circuit for all problems in this section, just changing the component values. For our first experiment, let's solve for total voltage. Total resistance in this circuit is 3000 ohms, and total current is 3 ma. or 0.003 amperes. By using the wheel, you can see that

$$Et = It \times Rt, \text{ or } Et = 0.003 \times 3000.$$

Putting these numbers into a calculator, you will find the battery voltage is 9 volts.

If we carry this a step further, we can calculate the voltage across each resistor. By using Kirchhoff's current law (which will be explained completely in the last section of this chapter) we can tell much more about a circuit. Briefly, Kirchhoff's law states that the total current in a series

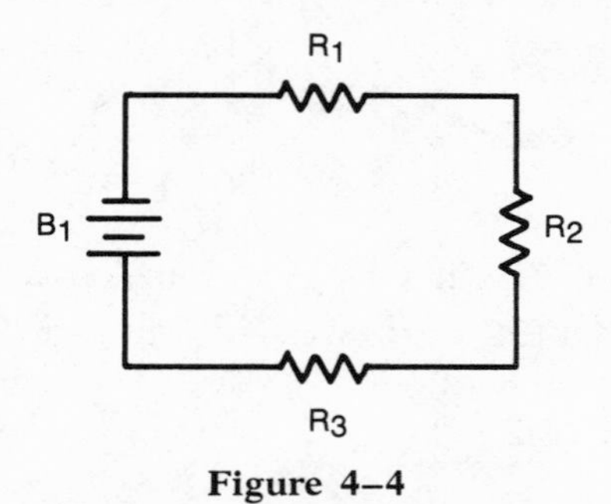

Figure 4–4
Series Circuit

circuit is the same in all parts of the circuit. This means that if the total current is 1 amp, then 1 amp will flow through each resistor. In Chapter 1, we saw this demonstrated when we learned how to measure current.

By applying this law, we can find the total voltage across each resistor. Using the wheel, and assuming R1 is a 1000-ohm resistor, we find that

$$Er1 = Ir1 \times R1, \text{ or } Er1 = 0.003 \times 1000;$$
$$\text{therefore } Er1 = 3 \text{ volts.}$$

Solving for other voltages around a circuit is just as easy. If you know the current of a component, and the resistance of that component, you may figure the voltage across it. One pitfall many students and experimenters run into is that they try to use the total voltage and the value of only one resistor in the circuit. When you do that, your answer will be wrong. Remember always to use the voltage and resistance that are equivalent; if you use total resistance, use total voltage.

Solving for current in a series circuit is easy. If you can calculate for either total current, or current through any resistor, you will obtain the current through the entire circuit. In our example circuit, let's make the battery a 6-volt unit. The three resistors will assume a total resistance of 6600 ohms. Total current in this case will be

$$It = Et/Rt, \text{ or } It = 6/6600,$$
$$\text{totaling } 0.0009 \text{ amps or } 0.9 \text{ ma.}$$

Solving for total resistance in a series circuit is also easy. If you know the values of the individual resistances, total resistance is the sum of the individual resistances. Logically, if we know the total resistance of the circuit in Figure 4–4, and also the values of two of the resistors, we can calculate the third. Let's try one. R1 and R2 are 2000 ohms, total resistance is 5000 ohms, what is the value of R3?

$$R3 = Rt - (R1 + R2).$$

Solving inside the parentheses first, R1 + R2 equals 4000 ohms. R3 = 5000 − 4000, as a result, R3 = 1000 ohms. Any resistance problem is easily solved in this manner.

Now that you have seen how to calculate any individual component or circuit parameter, let's try a few examples.

Table 4–3 lists several possible component values for the circuit in Figure 4–4. Take a sheet of paper and cover the solution half of the table, and solve for each of the problems. As you check for each value, lower the paper to reveal the solution for each problem. When you feel you can solve for any given value, continue with the next section as we learn multiple solving.

It is possible to solve for all values in a circuit, if you are given the right values to solve for. Using the same circuit, we will practice using the technique. When you have achieved this skill, you will be well on your way to understanding how simple circuit changes affect all parts of a circuit. This will help you in the future as you modify or change values in practical experimenter's circuits.

Individual Solving of Values in Figure 4–4

ET = 9 V	IT = 0.009 A	RT = ?	I = 0.02 A	R = 1 K	E = ?	
ET = 5 V	RT = 2 K	IT = ?	ER1 = 10 V	ER2 = 20 V	ER3 = 10 V	ET = ?
R1 = 2 K	R3 = 2 K	RT = 6K R2 = ?	R1 = 1 K	R2 = 2k	R3 = 5K	RT = ?

Solutions

9/0.009 = 1000 OHMS or 1 K	0.02 × 1000 = 20 VOLTS
5/2000 = 0.0025 AMPS	10 + 20 + 10 = 40 VOLTS
6000 − (2000 + 2000) = 2 K OHMS	1000 + 2000 + 5000 = 8 K

Table 4–3
Component Values for Figure 4–4

In our first example, we know the following values:

R1 = 1 Kohm; R2 = 3 Kohms; Rt = 6 Kohms; It = 10ma.

Solve for Et, Er1, Er2, Er3, Ir1, Ir2, Ir3, and R3. There is often more than one way to solve a given problem, so when looking at these values, don't expect that the first value will be the one you are looking for. The first step in solving this problem is in determining the value of R3.

$$R3 = Rt - (R1 + R2) = 2 \text{ Kohms.}$$

Next, solving for I is easy:

$$Ir1 = Ir2 = Ir3 = It \text{ (Kirchhoff's law)} = 10\text{ma.}$$

Solving for voltage:

$$Et = It \times Rt, \text{ or } Et = 6000 \times 0.01 = 60 \text{ volts.}$$

Solving for Er1:

$Er1 = Ir1 \times R1; Er1 = 0.01 \times 1000 = 10$ volts.

$Er2 = Ir2 \times R2; Er2 = 0.01 \times 3000 = 30$ volts.

$Er3 = Ir3 \times R3; Er3 = 0.01 \times 2000 = 20$ volts.

A quick check of accuracy is to apply Kirchhoff's voltage law. If you remember from Chapter 1, the sum of the voltages dropped in a circuit must equal the source voltage. In this case, Et = 10 + 20 + 30, therefore Et = 60 volts. This total matches the total calculated earlier, verifying our answer.

Table 4–4 contains multiple solving problems, again using the circuit of Figure 4–4. Using a sheet of paper, solve the problems just as you did in the previous table. When you feel you have a working ability with these calculations, begin the next section of this chapter. We will see how the rules you learned must be modified when working with parallel circuits.

Multiple Solving in Series Circuits
Use the Circuit in Figure 4–4

A. Rt=3 K Et=9 V R1=R2=R3
It=? Ir1=? Ir2=? Ir3=? Er1=? Er2=? Er3=?
R1=? R2=? R3=?

B. Ir1=1 ma R1=100 K R2=50 K R3=50 K
It=? Rt=? ER1=? ER2=? ER3=? Et=?

C. It=100 ma Er1=20 V Er2=30 V Er3=40 V
Ir1=? Ir2=? Ir3=? Et=? R1=? R2=?
R3=? Rt=?

D. Er1=20 V Ir1=.5 A Rt=160 ohms Er2=40 V
Et=? Er3=? It=? R1=? R2=? R3=?

SOLUTIONS

A. It=9/3000=3 ma Ir1, Ir2, Ir3=It=3 ma
Er1+Er2+Er3=Et Er1=3 V
Er2=3 V Er3=3 V R1+R2+R3=Rt
R1=1 K R2=1 K R3=1 K

B. Rt=100 K+50 K+50 K=200 K It=Ir1=1 ma Er1=.001×100 K=100 V
Er2=.001×50 K=50 V Er3=50 V Et=Er1+Er2+Er3=200 V

C. Ir1 = Ir2 = Ir3 = It = 100 ma Er1 + Er2 + Er3 = Et = 90 V
R1 = 20/.1 = 200 ohms
R2 = 30/.1 = 300 ohms R3 = 40/.1 = 400 ohms
Rt = 200 + 300 + 400 = 900 ohms

D. Et = 0.5 × 160 = 80 V Er3 = 80 − (40 + 20) = 20 V
It = Ir1 = .5 A
R1 = 20/0.5 = 40 ohms R2 = 40/0.5 = 80 ohms
R3 = 20/0.5 = 40 ohms

Table 4–4
Multiple Solving in Series Circuits

Solving for Voltage, Current, and Resistance in Parallel Circuits

Now that you have found out how easily you can analyze circuits using only a calculator and a schematic, let's see how to apply this knowledge to parallel circuits. In one respect, parallel and series circuits are opposites. As a quick reminder of the differences, let's review for a moment. While current is the same in all parts of a series circuit, voltage is the same in all parts of a parallel circuit. In series circuits, the sum of the voltages equals the source while the sum of the currents equals the source current in a parallel circuit. As you will recall, parallel circuits have all components connected across the source. Figure 4–5 diagrams the parallel circuit that we will use throughout this section. Calculating total voltage in a parallel circuit is easy if we know only the value of a single resistor, and the current through it. To calculate current in one branch, though, you must be able to figure the total current and the other

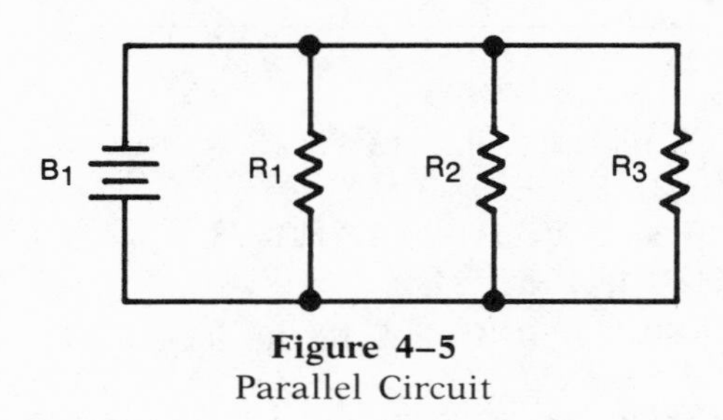

Figure 4–5
Parallel Circuit

currents; or you must know voltage and resistance of the branch. Total resistance, you will recall, is always smaller than the smallest resistance. You will learn some techniques for solving for total resistance without knowing current or voltage at the end of this section.

Let's start with a simple problem:

Et = 10 volts; R1 = 10 ohms; R2 = 10 ohms; R3 = 10 ohms;

Solve for Ir1, Ir2, Ir3, It, and Rt. Starting with Ir1:

Ir1 = Er1/R1;

or Ir1 = 10/10;

Ir1 = 1 amp.

As you can see, Ir2 and Ir3 can be calculated in the same manner. Total current is thus:

It = Ir1 + Ir2 + Ir3 = 3 amps.

Total resistance is calculated with the formula from the wheel R = E/I.

Rt = 10/3 = 3.3 ohms.

As you can see, total resistance is smaller than the smallest resistor, and the sum of the currents equals total current, therefore our answer checks.

Trying a more complex example:

It = 9 ma.;

R1 = 1000 ohms;

R2 = 1000 ohms;

R3 = 1000 ohms.

Calculate Et, Ir1, Ir2, Ir3.

Beginning with a little deductive reasoning, since all three resistors are the same value, we can draw the following conclusions. Equal currents will flow through each resistor, and these currents will be one-third of the total. In other words, each resistor is drawing 3 ma., (adding Ir1 + Ir2 + Ir3 = 9 ma.). Calculating voltage comes next in our example:

Er1 = Ir1 × R1;

Er1 = .003 × 1000;

Er1 = 3 volts.

Since all voltage is the same, we have calculated Er1, Er2, Er3, and Et in one operation.

Table 4–5 contains some sample problems with values assigned to components in Figure 4–5. When you feel you can solve for voltages and currents, we will cover solving for total resistance alone.

There are many times when you might like to know the total resistance of circuits without knowing any values of voltage or current. Again using Figure 4–5, lets see how this is done. There are three methods for calculating total resistance in a parallel circuit. Though they have been given various names over years, I have come to know them as the "equal branch" method, the "product-over-sum" method, and the "reciprocal" method. Each can be used in certain situations, and they are listed in order of difficulty. The equal branch method requires that all resistors in all branches have equal value. In the following example, our circuit contains three 27-Kohm resistors. To solve for total resistance, divide the value of one resistor by the number of resistors that are the same. In this case, 27000/3 = 9000

Solving for E and I in Parallel Circuits
Use the Circuit in Figure 4–5

A. Et = 10 V   R1 = 10 ohms   R2 = 20 ohms   R3 = 40 ohms
Ir1 = ?   Ir2 = ?   Ir3 = ?   It = ?

B. Ir1 = 2 A   R1 = 50 ohms   R2 = 100 ohms   R3 = 100 ohms
Et = ?   Ir2 = ?   Ir3 = ?   It = ?

C. R1 = 1 K   Ir2 = 100 ma   It = 300 ma   Et = 100 V
Ir1 = ?   Ir3 = ?   R2 = ?   R3 = ?

SOLUTIONS

A. Ir1 = 10/10 = 1 A   Ir2 = 10/20 = .5 A   Ir3 = 10/40 = .25 A
It = 1 + .5 + .25 = 1.75 A

B. Et = Er1 = 2 × 50 = 100 V   Ir2 = 100/100 = 1 A   It = 2 + 1 + 1 = 4 A

C. R1 = 100/1000 = 100 ma   Ir2 = 300 − (100 + 100) = 100 ma
R2 = 100/.1 = 1 K   R3 = 100/.1 = 1 K

Table 4–5
Solving for E and I in Parallel Circuits

ohms. Total resistance is 9000 ohms; prove it by calculating current for each branch assuming a 9-volt battery. The sum of the currents should equal the total current calculated when you solve for It using Et and Rt. If there are only two resistors in the circuit, you divide the value of resistance by two. Divide by the total number of resistors, whether there are two or ten. Don't fall into the trap (of many beginners) of first adding the resistances together, then dividing. If you do it that way, you will obtain an answer that is the same value as that of the resistors.

Though the equal branch method is the simplest, it can be used only in the rare situation when all resistors are equal. The product-over-sum method is more versatile in that it can be used to calculate the value of any two resistors in parallel. The name comes from the formula, which is why I like to refer to it as such. Another name for the same formula is the unequal branch method. Product-over-sum means this:

$$\frac{R1 \times R2 \text{ (product)}}{R1 + R2 \text{ (sum)}}$$

In an example circuit with two 1000-ohm resistors, let's apply the formula.

$$1000 \times 1000/1000 + 1000; \text{ or } 1{,}000{,}000/2000 = 500 \text{ ohms.}$$

We could have used the equal branch method here, and the problem would have come out correctly. The emphasis is on the fact that I could have used two resistors that did not have the same value, and still have gotten the correct answer. Do not try this method with three resistors (e.g. R1 × R2 × R3) because it will not work. To solve for three unequal resistors, the reciprocal method is required. The formula for this is as follows;

$$\frac{1}{\frac{1}{R1} + \frac{1}{R2} + \frac{1}{Rn}}$$

To use this method, if you have the reciprocal key on your calculator, you will have an easy time of it. Our example circuit contains three resistors; R1 = 1 K; R2 = 2 K and R3 = 3 K. Start by getting the reciprocal of each resistor.

For those who need a reminder, a reciprocal of a number is that number divided into 1. The reciprocal of R1 is 1/1000 or 0.001. The reciprocal of R2 is 1/2000 or 0.002, and R3 is 0.003. We now add up these values and our answer is 0.006. The last step in solving this problem is to get the reciprocal of 0.006, which is 166.7 ohms. This seems complicated, so we will try one more example to be sure you understand.

R1 = 10 K; R2 = 2 K; R3 = 5 K; what is Rt?

1/1000 = 0.0001; 1/2000 = 0.0005; 1/5000 = 0.0002,

and the sum of these is 0.0001 + 0.0005 + 0.0002 = 0.0008.

The reciprocal of 0.0008 = 1/0.0008 = 1250 ohms; Rt therefore equals 1250 ohms. Since the answer is smaller than the smallest resistor (2 K), we are probably correct. To double-check, substitute a battery voltage of your choosing, and solve for currents as previously.

Table 4–6 contains more problems on which to practice your newly found skills. Remember, any skill takes time and practice to develop proficiency, so spend a little time now and work the sample problems. Though many experimenters have gotten along with little experience in this area, they are limited to the simplest projects, and they never really understand what is going on in a circuit. Practice will bring understanding, and a deeper knowledge of the opera-

Solving for Resistance in Parallel Circuits
Use the Circuit in Figure 4–5

For problems A through D, pretend R3 is not in the circuit.

A. R1 = 10 K R2 = 10 K Rt = ?

B. R1 = 200 ohms R2 = 400 ohms Rt = ?

C. R1 = 27 K R2 = 27 K Rt = ?

D. R1 = 1 M R2 = 500 K RT = ?

Now, include R3.

E. R1 = 3 K R2 = 3 K R3 = 3 K Rt = ?

F. R1 = 10 K R2 = 20 K R3 = 30 K Rt = ?

G. R1 = 56 K R2 = 56 K R3 = 28 K Rt = ?

H. R1 = 12 K R2 = 24 K R3 = 36 K Rt = ?

SOLUTIONS

A. 10000/2 = 5 K

B. (200 × 400)/(200 + 400) = 133.3 ohms

C. 27000/2 = 15.5 K

D. (1000000 × 500000)/(1000000 + 500000) = .333 K

E. 3000/3 = 1 K

F. $$\frac{1}{\frac{1}{10000} + \frac{1}{20000} + \frac{1}{30000}} = 5.5 \text{ K}$$

G. $$\frac{1}{\frac{1}{56000} + \frac{1}{56000} + \frac{1}{28000}} = 14 \text{ K}$$

H. $$\frac{1}{\frac{1}{12000} + \frac{1}{24000} + \frac{1}{36000}} = 6.5 \text{ K}$$

Table 4–6
Solving for R in Parallel Circuits

tion of the electronic circuits you will build. Putting together series and parallel circuits is the last step. When you understand the processes and procedures in the next section, we will leave the math alone for awhile and proceed to more interesting topics.

Solving for Voltage, Current, and Resistance in Complex Circuits

We have worked with the two simplest types of circuits in the last sections. Unfortunately, only the least complex circuits that experimenters use are like those. The majority of experimenters' circuits are more complex in that they contain both series and parallel elements. You were introduced to these in the previous chapter, so let's see how you can solve for their respective voltage, currents, and resistance. It might be a good idea to review the last section on combination circuits in Chapter 3. When you feel confident that you can understand the operation of the complex circuit, return to this point and continue reading.

The rules learned earlier still apply—complex circuits do not modify any laws, and they must just be applied at the right time and in the right parts of the circuit. There are so many variations of complex circuits that it would be impossible to outline each one. The objective here will be to gain an understanding of how to apply those rules learned up to this point. Remembering the two general rules of complex circuits will help. For review they are: (1) Treat all series connected components using series rules, and (2) Treat all parallel connected components using parallel rules.

Refer to Figure 4–6 for the following circuit descriptions as we solve for total resistance in the circuit. When you solve for complex circuits, follow these simple steps.

1. Solve for series resistances that are within parallel branches. An example of this type is R3 and R4 in the figure.
2. Solve for parallel branches next. In our example, this is R2 in parallel with the combination R3 and R4.
3. Solve for the series components last. In this case, R1 and the solution of R2 with R3 and R4.

Let's try to solve the problem following the rules. All resistors are 1000 ohms. **Step 1:** Solve for series parts of parallel branches. In this case, R3 + R4 = 2000 ohms. **Step 2:** Solve for parallel branches. Product-over-sum will work best here. Use

$$\frac{1000 \times 2000}{1000 + 2000}$$

The solution equals 667 ohms (approximately). **Step 3:** Solve the series parts last.

$$\text{R1} + (\text{solution to parallel branch}) = \text{Rt.}$$
$$\text{Rt} = 1000 + 667 = 1667 \text{ ohms.}$$

Solving for resistance is simply a matter of breaking down the program to its least complex parts, and solving it internally, working towards the outside.

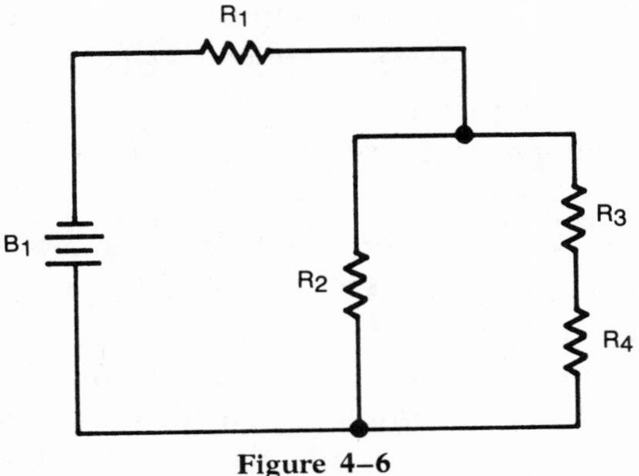

Figure 4–6
Combination Circuit

Analyzing for voltage and current is slightly more difficult. In our example, there is only one path for current to flow from the negative battery pole, through R1. Total current must travel through R1, and a voltage drop will appear across it. The current divides at the junction of R2 and R3, and recombines at the junction of R2 and R4. Each branch will drop the remaining voltage not dropped across R1. Voltage drops are the same in a parallel circuit, so the voltage drop that appears across R2, *or* the voltage drop that appears across R3 and R4 will add to the drop across R1 and equal the total voltage in the circuit. The two branch currents will add, their sum being equal to the supply current. Notice that this does not mean the two currents are equal; they will be equal only if total resistance in each branch will be equal.

Let's apply 10 volts to the test circuit in Figure 4–6 and see what happens. After figuring the total resistance at 1667 ohms, total current is easy.

$$I = E/R \text{ or } I = 10/1667 = 6 \text{ ma., approximately.}$$

If total current is 6 ma., then current through R1 must also be 6 ma. Now we can figure the total voltage drop across R1.

$$Er1 = Ir1 \times R1 = 0.006 \times 1000 = 6 \text{ volts.}$$

The voltage drop across the parallel branch must then be 4 volts, again because the sum of the drops equal the source voltage.

We can now determine the currents in each branch.

$$Ir2 = Er2/R2 = 4/1000 = 0.004a.$$

The current through the series resistors R3 and R4 can be solved by solving for total branch current. Branch current in this case is equal to:

$$Ibr = Ebr/Rbr = 4/2000 = 2 \text{ ma.}$$

We could have determined this also by just subtracting the other branch current from total branch current, as the sum must equal 6 ma. As you can see, solving complex circuits is similar to solving any puzzle. You must determine which values you know and use those values to solve for things you don't. As you solve for more parameters, you gain more clues to the total solution to the problem. Try a few on your own; Table 4–7 contains some experimental values to be applied to the circuit in Figure 4–6. As you have done in the past, take a sheet of paper and cover the solutions to the problems, then check them after you have worked the problem.

Solving for Values in Complex Circuits
Use the Circuit in Figure 4–6

A. Et = 100 V R1 = 150 ohms R2 = 300 ohms R3 = 150 ohms R4 = 150 ohms
   Rt = ? Er1 = ? Er2 = ? Er3 = ? Er4 = ?

B. Er1 = 20 V It = .2 A Er2 = 20 V Ir3 = .1 A R3 = 100 ohms
   R1 = ? R2 = ? R4 = ? Et = ? Er3 = ? Er4 = ?

C. Et = 60 V Er2 = 30 V Er3 = 20 V It = 10 ma Ir2 = 5 ma
   R1 = ? R2 = ? R3 = ? R4 = ? Ir4 = ? Er4 = ?

SOLUTIONS

A. Rbr = (150 + 150) = 300/2 = 150 ohms Rt = 150 + 150 = 300 ohms
   Er1 = 100 − 50 = 50 V Er2 = 100 − 50 = 50 V Er3 = 50/2 = 25 V
   Er4 = 50/2 = 25 V

   You could have solved this problem by solving for current, rather than using Kirchhoff's voltage law as I did. The results would be the same.

   There are many variations for solving these problems. Solve them any way you can. Even if your formulas and steps are different, you should have obtained the same answers.

B. R1 = 20/.2 = 100 ohms   Ir2 = Ir3 − Ir1 = .1 A
   R2 = 20/.1 = 200 ohms
   Er3 = .1 × 100 = 10 V   Er4 = 20 − 10 = 10 V

C. Er1 = 60 − 30 = 30 V   R1 = 30/.01 = 3000 ohms
   R2 = 30/.005 = 6000 ohms
   Ir4 = It − Ir2 = 5 ma   R3 = 20/.005 = 4000 ohms
   Er4 = Er2 − Er3 = 10 V
   R4 = 10/.005 = 2000 ohms

Table 4–7
Solving for Values in Complex Circuits

Kirchhoff's Law Finally Understood

Students of electronics have often been mystified by Kirchhoff's laws. Let's review them, and you will see that they are really descriptions of the behavior of circuits under potential. When you understand and apply these laws, you will find them useful whenever you attempt to analyze how a circuit operates.

We have applied Kirchhoff's voltage law many times already. Simply stated, the sum of the voltage drops, in a series circuit, equals the source voltage. Whenever you use a series circuit, you can see this property in use, from the simplest two-bulb circuit, to a complex string of transistors stacked in series to drive a set of speakers. Knowing the source voltage, you can determine the drops across each device, or knowing the drops will allow you to determine the source. Applying this to a real circuit (use Figure 4–2), suppose the LED doesn't light, even though we have applied the correct voltage. We know the circuit should draw about 20 ma., and the voltage across it should be 1.2 volts. Measuring the voltage across the resistor, we find that it is zero. Measuring across the LED, the entire supply voltage appears. Kirchhoff's law will tell us that the LED is defective, and open. One might come to the conclusion that the resistor is shorted; however, that is quite unlikely, and if it were likely, the LED would probably open from excess current flow in any case. If you suspect a problem with the resistor, inspect it. Shorted resistors usually will look as if they have been warm, and have a charred appearance.

Kirchhoff's current law is another example of performance of electricity in a circuit. Kirchhoff found that the number of electrons entering a point in a series circuit will be equal to the number of electrons leaving that point. An analogy of a full bus might apply. People entering the front of the bus must equal people leaving at the rear exit. In other words, we have another way of explaining that total current in a series circuit is always the same.

By applying these laws you can easily determine the voltage and current characteristics of any circuit. Keep these two concepts in mind; they will be referred to many times, especially as transistor circuitry is explained.

Refer to these two chapters as necessary if you need a review of how circuits can be analyzed. In the meantime, let's get back to some hardware. A short break from theory is always welcome. In the next chapter, we will look at tools and equipment the modern experimenter uses. New techniques will be explained and, as we describe the tools, you will be able to determine which ones will best fit your needs.

Accomplishing More With Wiring Tools And Techniques

5

The experimenter or technician needs several different hand tools, and several different types of soldering equipment, as well as test equipment and components. This chapter will outline commonly used hand tools, tools that must be acquired if you want effective results in wiring and building projects. Several tools that are nice to have, though not necessary, are also discussed. These tools will make the experimenter's work go faster and last longer.

Soldering techniques are also covered, and, after reading and practice, you will be able to solder with the professionals. If you have ever tried to remove components, or disassemble circuits that have already been soldered, you will appreciate the section on desoldering techniques. You will learn how professionals desolder circuits for repair, and you will learn some advanced techniques for desoldering.

A rapidly increasing number of experimenters are using wire-wrapping techniques instead of, or in addition to, soldering. The techniques, and the advantages and disadvantages, of this relatively recent addition to the experimenter's bench are explored and explained. The last section of this chapter discusses another relatively new technique for circuit prototype and repair. If you have ever spent hours meticulously soldering and constructing a circuit, only to find it doesn't work, you will appreciate the advantages of the prototype breadboard, or protoboard.

Hand Tools for the Experimenter

The list of required tools for the electronics experimenter is actually quite small. However, there are many tools that, though not expressly necessary, are quite useful in many applications. We will look at both groups, and you will be able to decide which tools you will need, and be able to select quality tools.

The needle-nose pliers, diagonal cutters, a standard screwdriver and a Phillips screwdriver constitute the minimum requirements for hand tools. Many small projects can be built using only these tools and a soldering iron. The soldering iron is discussed separately in the next section; it is not being classed as a hand tool for purposes of this discussion. Choosing quality tools here is particularly important, as these tools are used most often.

When purchasing tools, buy from a reputable dealer, who will allow you to return a tool that does not operate to your satisfaction. Remember, though, no one should have to accept a returned tool that clearly has been abused. For example, a screwdriver with a bent shaft has clearly been abused; it should never be used as a prybar. Bargain tools should be avoided, as they tend to get out of alignment or break when stressed, sometimes in the normal use of the tool. Figure 5–1 is a photograph of the standard tools required. Don't forget electronic service shops, which may have good quality used tools available. Inspection of the tool before purchase is probably the most reliable method of evaluation. Manufacturer's reputation, recommendation of friends, and cost should also be taken into consideration.

Taking each tool separately, we will see what you should look for. The needle-nose pliers are a real work horse for the experimenter. There are several different types of pliers available. For typical project work, consider six-inch pliers with insulated vinyl handles. Four-inch pliers are recommended for working in miniature circuits, as you might be,

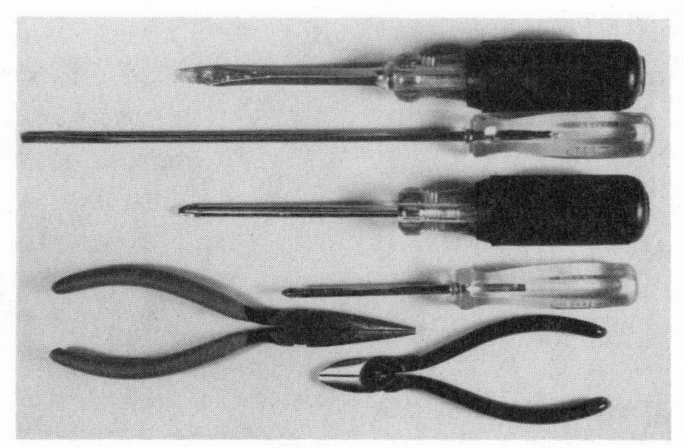

Figure 5–1
Commonly Used Hand Tools

when building IC projects. Some pliers contain built-in wire cutters in their jaws. Try to find pliers with this option if you can. When wiring, it is often convenient not to have to change tools just to shorten the wire. A quick snip with the cutter blade will allow you to continue working. When in the store, examine the jaws of a prospective purchase first. Point the nose of the pliers at you and make sure the tips of the jaws are in line. See Figure 5–2. Look at the serrations inside each jaw; they should be lined up exactly parallel, with no rough edges or uneven grooves. Work the pliers open and closed several times, they should move freely and easily without binding. The cutters should be sharp, and, when you hold the pliers up to a light, you should see no gap between the cutter blades.

The diagonal cutters, or dikes, are another necessary

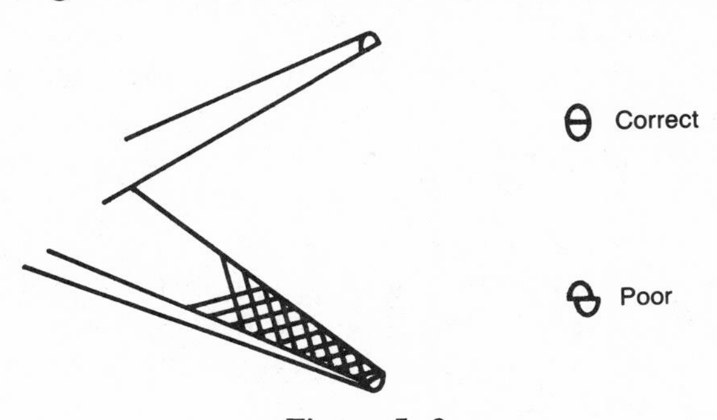

Figure 5–2
Illustration of Pliers Jaw

tool for the experimenter. They are used most often for clipping the lead ends from soldered connections. Though cutters are available in many sizes, the size most needed by experimenters is six-inch. As with the needle-nose, four-inch cutters are useful for miniature circuit wiring. Apply all the rules for testing needle-nose pliers to testing diagonal cutters. Don't use four-inch cutters to cut heavy gauge wire. The finely constructed jaws cannot take the pressure required to cut large wires, and the jaws will almost certainly break. Again, select a model with vinyl insulated handles.

Everybody has a common screwdriver around the house somewhere. There is probably no greater variation in tool quality than is found in the screwdriver. Prices, and quality, range from the 99-cent grocery store variety, to several dollars for a high-quality tool. The experimenter will require several sizes, including #1 and #2 point Phillips screwdrivers, and 3/16-inch and 1/4-inch flat-blade screwdrivers. These drivers should have a 6-inch shaft and insulated handles. Some projects will require "stubby," or 2-inch shaft drivers as well. You can sometimes save some money by purchasing a set of drivers. Be especially cautious of quality when choosing a set, and insist on being able to inspect each tool for quality before buying. Sets are also available as different blade and shaft assemblies, with an interchangeable insulated handle. This is the least expensive way to acquire a set, and is recommended for all but the most serious experimenter. You will take extra time exchanging blades in the handle, and, if you lose the handle, you cannot use any of the blades. Make sure that replacement handles and blades are available for any interchangeable screwdriver set.

When purchasing these tools, look for smooth, polished corners on the tips of drivers. Hold the driver in your hand and feel for rough edges in the plastic. Phillips drivers should have no flaws on their tips; compare the tips of the most expensive drivers to those of the one you are examining. If you buy an interchangeable handle set, insert several blades into the handle. They should insert with moderate force and lock into place. Grab the handle with one hand,

and the blade with the other. Apply pressure to the driver as if trying to break it in two. There should be little or no play, or side-to-side motion, when doing this. Screwdrivers with ratchet handles or ball grips are sometimes handy, though they would definitely fall into the group of nice-to-have tools, rather than needed ones. You would probably be better off investing in a higher-quality set, with more blade styles, then purchasing a ratchet driver. The major advantage of the ball ratchet driver is not having to remove your hand when wanting to apply torque for another turn of the driver. The ball also is easy to grip, and multiplies the torque your hand applies. The major disadvantage of the ball is that it is unusable in miniature circuits and tight places.

Let's take a look at some of the tools that are nice to have, though not really necessary. Figure 5–3 illustrates some of these tools. The most commonly found tool in this category is the wire stripper. It is available in many forms, though the most typical form is in a pliers-like device that has a cutter's end with a small hole where the wire is placed. To operate the device, set the gauge wheel to the proper wire gauge for the wire you are using. Insert the wire in the jaws, clamp the strippers on the wire, and, while holding on to the wire, pull the strippers away from you. The cutters will sever the insulation without cutting the wire. This tool is much easier to use than a knife for removing insulation from wires. I had trouble deciding whether it should be in the need-to-

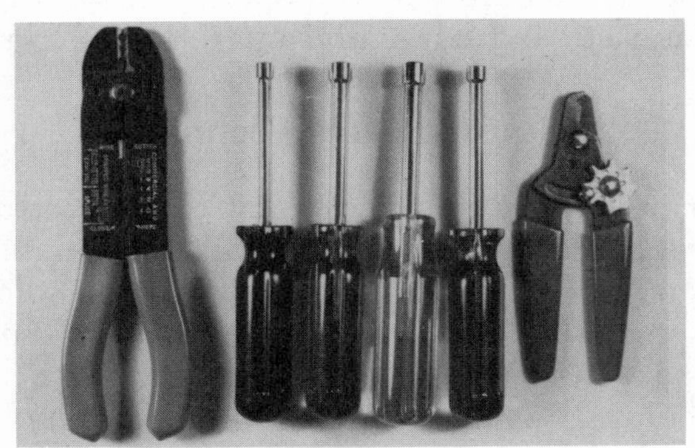

Figure 5–3
Typical Nice-to-Have Tools

have list, but listed it here because it is more difficult to strip a wire without one. One caution—be sure not to nick the wire with the strippers by selecting too small an opening. If the wire doesn't immediately break, it will certainly break when stressed. Nothing is more frustrating than to have a wire break off just after you have finished soldering it to a terminal.

Another handy tool to have is the nut driver. It is built like a screwdriver, but has a hex socket on the end of a hollow shaft. This allows the tool to slip over the end of a bolt, where the hex head can contact the nut. Turning the driver will loosen or tighten the nut. These drivers are available in several sizes, but the most commonly used sizes are 1/4-inch and 5/16-inch for typical electronic hardware. You may purchase them in complete sets or individually.

Another convenience device is the six-in-one tool, which contains a crimping tool to install solderless connectors on cables and wires, a wire stripper, small wire cutters, a wire size gauge, a small bolt size gauge and small bolt cutters. This is a very handy tool to have if you are a serious experimenter. It has so many useful features that it is well worth its cost.

There are many other useful tools that can be used by the experimenter. As you become more involved with experimenting, you will no doubt see tools that you might be interested in purchasing. Always use care when purchasing tools; check each tool for smooth operation, quality machining, precision cutting edge, and guarantee of return.

Basic Soldering Techniques

Soldering procedures are older than electronics. For many years people have used soldering techniques for joining two pieces of metal. There are many grades of solder, and it has many purposes. The stained glass windows in churches have had panes of glass installed between lead tracks that were meticulously soldered together. The automobile radiator uses a soldering process to keep the antifreeze solution from leaking. Let's take a closer look at electronic soldering and see not only what it is, but what it isn't.

Soldering in electronics has two basic purposes: to allow a tight, immobile connection in wiring, and to keep the surface from oxidizing, thus developing a low-resistance connection. Electronics soldering, indeed any soldering, is not done for mechanical strength. If you need mechanical strength, you need to weld the materials together.

Solder is not meant to join pieces of aluminum; therefore you may not solder any wires or terminals to aluminum chassis or boxes. Solder will join copper-plated and tin-plated wires to other wires, terminals, or printed circuit boards.

Solder is composed of an alloy, or a combination of tin and lead. This mixture determines primarily the application of the solder. I will discuss only the type of solder that is available for electronic purposes. Basically, solder is available with a mixture of approximately 60-percent lead and 40-percent tin. The melting temperature of 60/40 solder is 190 degrees C or 373 F. Varying the ratio of tin to lead will vary the melting point of the solder, and other characteristics, so be sure to specify solder near the 60/40 ratio. Solder is usually available on rolls, and comes in different gauges, or wire sizes; the most commonly used size for the experimenter is 18-gauge.

Solder works by dissolving and attaching to the surface of the material being soldered. In order for the fusion process to work, the surfaces to be soldered must be clean and free of oxidation. The purpose of soldering flux is to clean, deoxidize, and prepare the metal to accept a coating or layer of solder. Some metals requiring industrial soldering need a very strong flux. This acid flux is extremely corrosive and will cause electronic connections to deteriorate over a short period of time. Never, repeat **NEVER,** use acid-core solder in an electronic project. Within a short period of time the wiring connections will literally fall apart. Acid fluxes are sometimes conductive as well, and could cause unwanted current paths to flow in a circuit.

Electronic soldering requires a rosin flux, which is an inactive chemical at room temperature, but when heated becomes active, cleaning and preparing the metals for joining. Unlike industrial metal soldering, electronic flux is enclosed in the core of the solder you purchase. There is no paste flux that must be brushed on the connection to allow it to work; the action of melting the solder releases the flux and allows it to clean and prepare the connection.

When buying solder for electronic purposes, order 60/40, 18-gauge, rosin-core solder. If you do a lot of building, you might want to buy a one-pound roll, however many experimenters buy smaller quantities, due to the relatively high cost of solder.

As was stated earlier, the melting temperature of solder is around 373°F. The soldering iron or gun is the tool usually used to melt the solder. The ability of a gun or iron to deliver heat is measured in watts; a small iron typically dissipates 25 to 40 watts, while a larger iron or gun will dissipate 100 watts or more. Before we discuss the actual soldering procedure, let's look at the tools available and their advantages and disadvantages.

Soldering guns are usually pistol-shaped, and include a "trigger" that turns the gun on when pressed. There are many styles available. See Figure 5–4 for an example of a typical gun. The advantages of the gun are rapid warm-up and ability to deliver high heat to large electrical connections. The disadvantages of the gun are its relatively large size and correspondingly large tip, which make soldering transistors and other small components all but impossible. Its high heat level can easily destroy integrated circuits and can be more of a hindrance than a help.

For soldering small transistor and integrated circuit projects, the iron is the recommended tool. Choose an iron no larger than 40 watts, and a 25-watt iron is sufficient for most applications.

A good quality iron will have many tip sizes available, and they will be easily replaceable. A 1/16-inch chisel tip is usually the style that comes with most irons when they are purchased, and it will suffice in almost all situations. If you plan on building projects that contain MOS devices, be sure the iron you select has a three-prong plug and do not use an adapter plug to a two-prong outlet. The ground pin on a three-prong iron is connected to the tip, and dissipates any

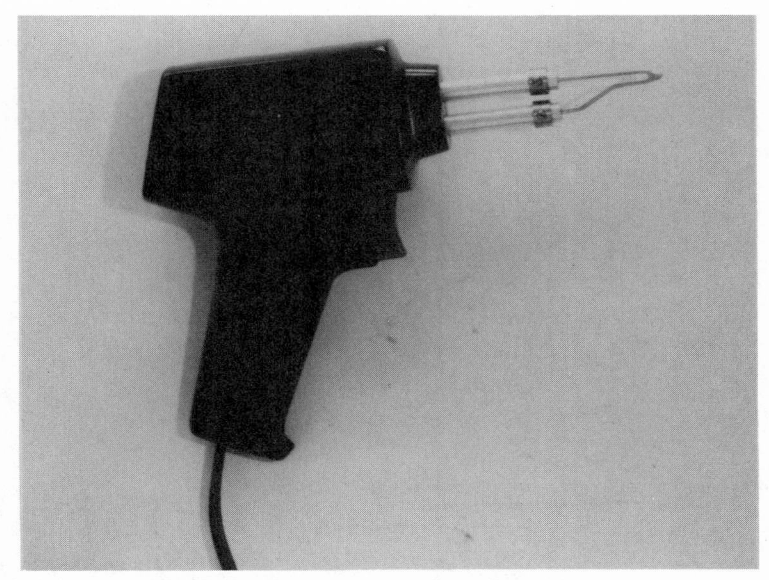

Photo courtesy of Radio Shack, a Division of Tandy Corp.

Figure 5–4
A Typical Pistol-Grip Soldering Gun

leakage current that might destroy the device being soldered.

Figure 5–5 illustrates typical soldering equipment. A nice touch, if you can afford it, is a thermostatically controlled iron. These irons have special tips that actually contain a thermostat, and cycle on and off as the temperature demands require. These irons are meant for continuous duty and will operate for long periods without excess wear on the tip. If you do not have one of these irons, do not keep your iron unused but plugged in for long periods of time. Tips treated in this manner will need to be replaced far sooner than is otherwise normal. Other hints for maximum iron tip life will be pointed out as you learn how to solder.

Since most of the projects in this book use transistors and integrated circuits, it is assumed that you have obtained a 25-to-40-watt pencil-style iron. Though you could purchase an iron with a conical tip, or several different sizes of chisel

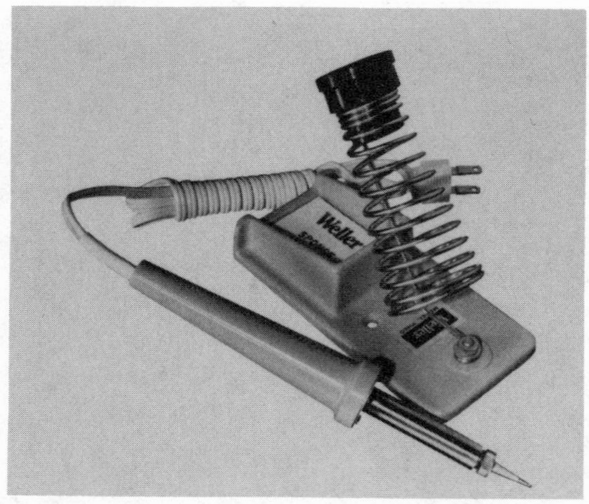

Photo courtesy of Weller Industries, a division of the Cooper Group.

Figure 5–5
A 25-Watt Soldering Iron with Stand

tips, if you are planning on purchasing only one or two tip styles, the 1/16-inch chisel tip is nice for most work, with a 1/32-inch conical tip for use when soldering integrated circuits.

While at the store, you should pick up some 60/40 18-gauge solder, and a roll of desolder wick braid. Another useful accessory is a metal soldering iron stand, which will hold the hot iron while it is not being used. The section on desoldering techniques will also mention some "nice-to-have" accessories that you may wish to purchase.

The first step in soldering is finding something to solder. If you wish to practice, and don't have a junk box full of parts to practice on, you can visit a TV shop or an electronics flea market, where you can get a junk circuit board or two to practice on. Don't pay too much for one of these; indeed, most TV shops will give you an old TV chassis to experiment with, if you handle the disposal problem for them. Since, in this case, you are not going to be soldering any new components to the circuit, skip ahead to the desoldering section, where you can learn those techniques first. Remove several components from the chassis, or circuit board, then practice soldering them.

Enough theory—let's get down to soldering techniques.

The first step is to clean and tin the iron. After the iron is warm, wipe the tip on a damp cloth or sponge. This removes oxidation, old solder, and excess flux from the tip and readies it for tinning. Proper cleaning of the tip not only assists in making a solid connection, it actually prolongs tip life. Apply a small amount of solder to the tip of the iron and allow it to melt. Be careful not to use too much; the most common mistake made by students is using too much solder. The solder should flow evenly around the tip. If the tip is a chisel point, turn it over and tin the other side. A conical tip will usually allow the solder to flow around all sides. If the solder flows into a little ball rather than flowing smoothly, you have probably used too much solder, or the iron tip is still dirty. Wipe the excess solder from the tip and examine the tip carefully. If it is corroded or pitted, it may require filing. Tips that are properly cared for seldom require filing, and if you think yours does, it is best to refer to the manufacturer's literature on cleaning and filing the iron. Some tips, especially on temperature-controlled irons must never be filed. Use an old file, as the solder and light filings will soon plug and ruin a good file.

Now that we have a clean and tinned iron, the next step is to heat the connection. There are two commonly made connections that the experimenter usually solders. These are component leads to terminals or connectors, and component leads to printed circuit boards. For these exercises, refer to the series of photographs of Figure 5–6 that depict the soldering process. To solder to a terminal strip, mechanically wrap the component lead around the terminal. Remember that soldering does not provide a lot of mechanical strength, especially on wires and cables that are flexed. Apply the iron to the connection, allowing the iron to touch both the leads being soldered and the terminal. After the connection has had a chance to warm, touch the solder to the point where the iron, wires, and terminal touch. Another common mistake that beginners make is to melt the solder on the iron. When this happens, it just runs on to the terminal and creates a "cold" solder joint. A cold solder joint is a high-resistance connection, and is especially subject to problems when the completed unit is moved or vibrated.

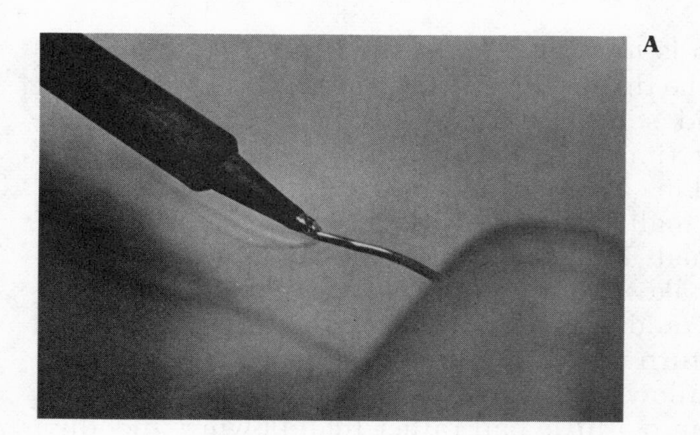

A

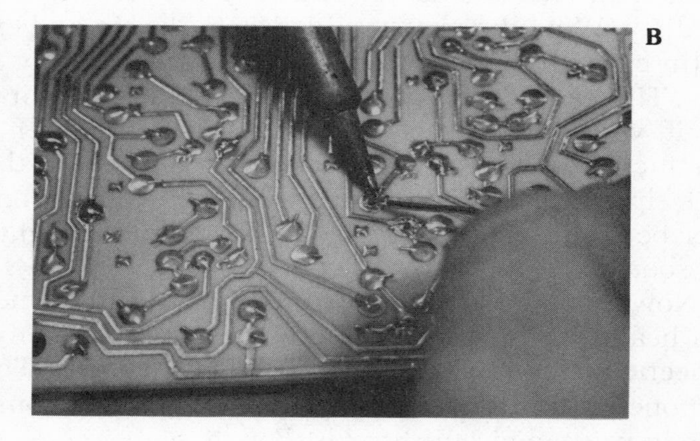

B

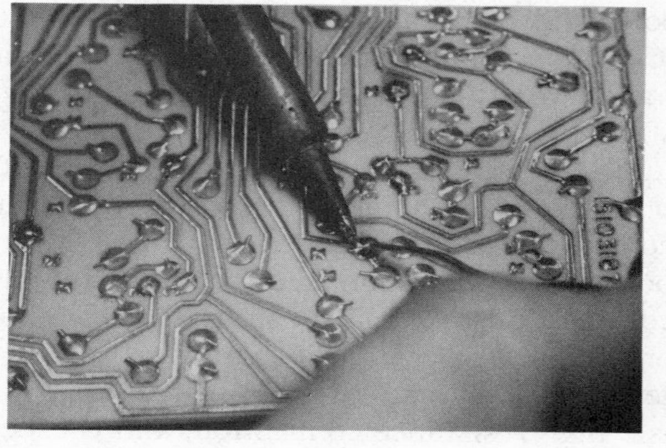

C

D

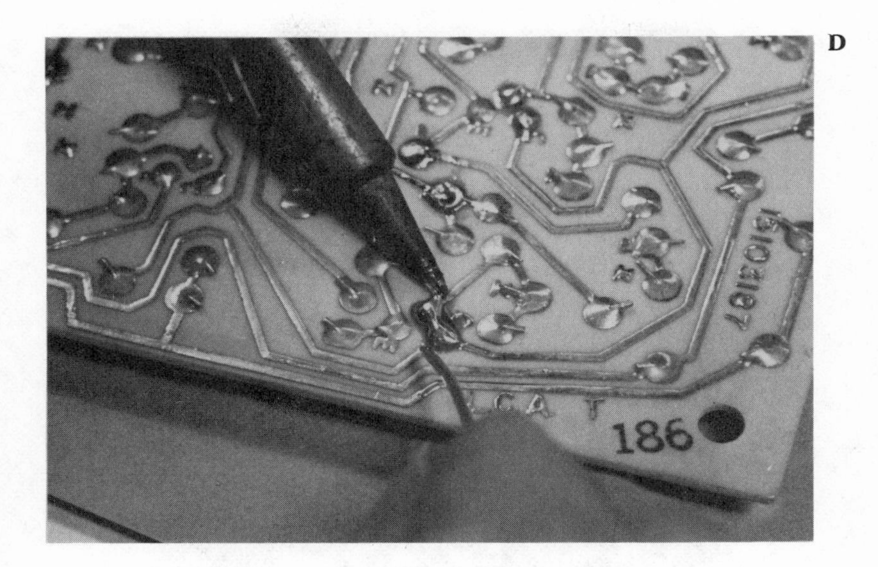

E

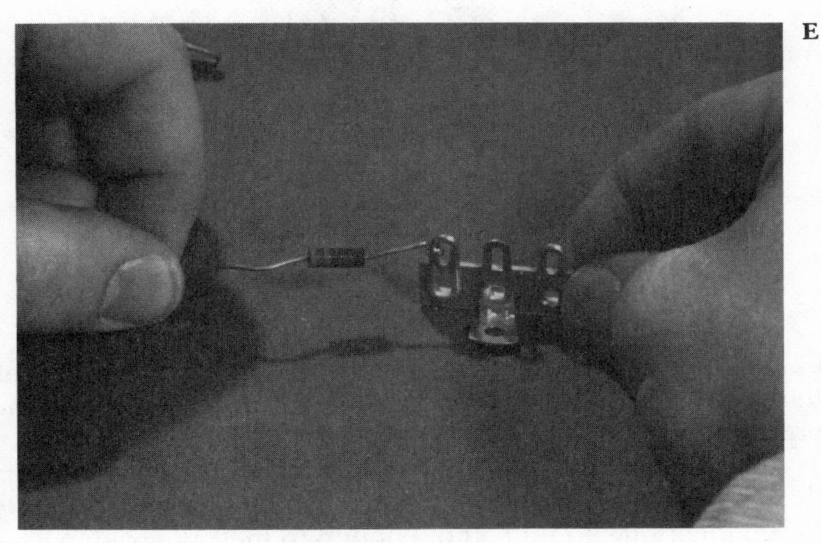

Figure 5–6
A. Tin the Soldering Iron.
B. Apply Iron to Connection, Contact Lead, and Foil.
C. Apply Solder to Connection.
D. Too Much Solder Can Create a Solder Bridge.
E. Good Connection on Terminal Strip.

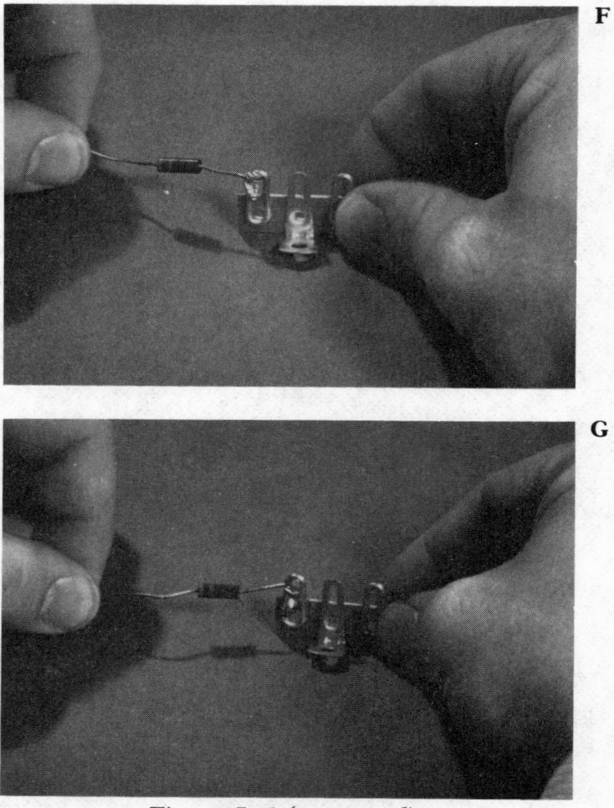

Figure 5–6 (continued)
F. Cold Solder Joint Caused by Component Movement
G. Too Much Solder on Connection.

Cold solder connections are responsible for well over half of the problems experimenters have when building kits or projects.

Now that you have touched the solder to the junction on the terminal, remove the iron. If you have soldered properly, the connection should be hot enough to melt the solder. The solder should flow smoothly and easily over the wires. If the solder balls, you have used too much, the connection is not clean, or the connection was not hot enough. Do not allow the wires to move about as the solder is hardening, or you will have a high-resistance connection. Inspect all wires and be sure they are soldered when you are finished.

Soldering to a printed circuit board is quite similar to soldering a terminal, but it is easier. Usually there is only a single wire that must be soldered to a single foil on the board. Again apply the iron to the junction of the lead and foil. Allow them to heat, and apply the solder to the junction of the wire, foil, and iron. One caution: Circuit boards heat up rather quickly, and application of excess solder could have serious consequences. The solder may flow along the copper and cover several component mounting holes. Before you continue building, you will have to open all of the plugged holes on the circuit board. This is a problem especially with integrated circuit projects, due to the fact that the holes are so small and closely spaced.

Another problem with excess heat is that the foil on the circuit board is held to the board by an adhesive. Though the bond between board and foil is relatively sturdy, heat can soften the adhesive, and the foil can lift from the board, causing some really serious problems. The section on desoldering contains more on this, as the problem occurs more often during the desoldering procedure.

The final test for a solder connection after it hardens is to flex the wire or component lead while watching the solder joint carefully. Any movement of the lead underneath the solder, or instability of the connection, can usually be cured by just reheating the connection. If there is not enough solder there to make a tight connection, apply a little more. Again, the usual problem is too much solder, rather than too little.

Learning to solder is easy; it just takes practice, as do all skills. Spend time watching the soldering process while you are practicing. You will soon learn the correct amounts of heat and solder to apply. Now to take up the process of desoldering, that is, removing those components we install—safely, and without damage.

Techniques of Desoldering

An important, though often overlooked, skill for the experimenter is the art of desoldering connections. At first, you might wonder why anything needs to be desoldered.

Just a couple of examples should demonstrate why. Much electronic equipment that has outlived its usefulness in its originally manufactured form can be disassembled and used for more current projects. Another time when desoldering is called for is when the experimenter accidentally solders a component into the wrong hole or terminal connection, or a defective part needs replacing.

Let's take a look at the three most popular methods of desoldering by the experimenter. Each will be covered individually in this section. The most common technique is the use of desolder braid, or solder wick. The desolder bulb is also often used, as is the "solder sucker" or desoldering tool. The braid is the least expensive initially, though since it is consumable, it must be restocked as it is used up. The solder sucker is the most expensive, with a purchase price of between ten and twenty dollars, depending upon the quality of the device. Industrial electronics technicians use solder vacuums that connect the test bench to a vacuum system. Melted solder is literally vacuumed off the board and into the desolder vacuum. This equipment is quite expensive, and very few experimenters have access to it.

The desolder braid is easy to use and is recommended for all but the most tedious and lengthy desoldering jobs. It is usually found in little plastic rolls only a few feet in length, and consists of a copper braid, sometimes treated with chemicals, that aids in the desolder process.

Desoldering with the braid is easy, especially with printed circuit boards. Removing integrated circuits is nearly impossible without using proper techniques, and is simple using the skills you will now be shown. Refer to Figure 5–7 for photos of the process. Start by identifying the leads of the component you will be removing from the circuit board. Place the end of the braid on the connection, and hold the braid by its plastic reel. If you try to hold the braid itself, you will burn your fingers, as the iron heats the braid. Next, place the tip of the hot iron over the braid. In other words, sandwich the braid between the connection and the hot iron. Allow the iron to heat the connection and the braid, and after a few moments, the excess solder softens

A

B

Figure 5–7
A. Using the Desolder Bulb
B. Using Solder Wick

and is drawn into the braid via capillary action. Remove the iron and the braid and inspect the connection. You will find that it has much less solder than it did before. The objective is to remove enough solder to break the connecting lead away from the edges of the hole. This may take a couple of applications of wick, especially if the original connection is large, or has too much solder on it. When you have released the lead from its connecting hole, it can be easily lifted out with needle-nose pliers. Take diagonal cutters and clip off the braid at the point where the solder ends.

The one problem you must watch for is applying too much heat to the printed circuit board. Integrated circuits and transistors are easily damaged by excess heat, and can be protected from the heat of the iron by clamping needle-nose pliers between the device and the lead being soldered. A worse problem is that excess heat may break down the adhesive that fastens the foil to the circuit board base. The foil will be damaged long before most modern electronic components would. Practice the desoldering technique on old circuit boards. Unsolder as many components as it takes for you to feel proficient in the desoldering process before attempting to remove an expensive integrated circuit.

Another common desoldering tool is the desolder bulb. This inexpensive tool is similar to an eye dropper. It has a stubby appearance, and consists of two parts, a tip made of space-age plastic that will not burn or melt at ordinary soldering temperatures, and a rubber bulb. Operating the device is simple. First heat the connection to be desoldered with the iron, and when the solder is soft it is ready to be removed. Squeeze the bulb hard to remove as much air as possible, then bring the tip of the bulb down to the connection. Remove the iron, and before the solder re-hardens, place the bulb on the connection and release the pressure on the bulb. The vacuum created when the bulb is released will pull the solder into the bulb. A couple of reheats on the connection will remove enough of the solder to allow removal of the component.

There are a couple of problems that can be associated with the bulb. It is relatively easy to plug the tip with solder, which often requires removal of the tip and insertion of a blunt pointed object to force the solder out of the tip. Also, the bulb must occasionally be removed from the tip and emptied, thus removing old bits of solder from the inside of the bulb. Using the bulb is relatively easy, but I personally like the wick method as it seems to do a little better job with less heat. The vacuum created by the bulb filling with air is not great enough to pull large amounts of solder away from the terminal, and the process seems to be slower than working with the wick.

The desoldering tool is similar to the desoldering bulb. Its major difference lies in the fact that the tool has greater vacuum, which allows for quicker removal of excess solder. The tool is usually an eight-inch to ten-inch-long cylinder that has a tip similar in construction to the tip on the bulb. The other end of the cylinder usually has a plunger, and a trigger of some sort. If you do a lot of desoldering, this tool is well worth the money. Its operation is simple and the tool requires little maintenance. After loading the tool, (usually this is accomplished by pushing the plunger down until it locks) and melting the solder on the connection, just remove the iron, quickly place the tip of the desoldering tool over the connection and press the trigger. The spring-loaded plunger is released and travels quickly up the cylinder. The vacuum created by the tool is much larger than that created by the bulb, and usually only one or two attempts will remove enough solder to allow the removal of the component. Maintenance of the iron is easy. The tip is removed, and the inside of the cylinder collects the solder, which usually sticks to the top of the plunger. Removal and reassembly are simple, and the only other items to be checked are the rubber O rings that seal around the edge of the plunger. As these rings age, the vacuum decreases, thus causing the tool to become inefficient. If a light coating of oil doesn't improve the performance of the seal, it will have to be replaced.

Solder has been used in equipment for years, however some new techniques have been developed that reduce or eliminate the need for soldering. The techniques are all based upon the wire-wrap tool. In practice, circuits that are wire-wrapped are as reliable and effective as those that are soldered, yet they are simple and easy to modify. Let's take a look at the process.

Wire Wrapping and Perf Boards

Wire wrapping is a technique that was developed partially because of the integrated circuit technology, and the relative complexity of the construction of printed circuit boards. Hobbyists, engineers, technicians, and experimenters needed a technique where they could simply interconnect components for building one-of-a-kind circuits. Industry was looking for a method that did not require all of the hassles of soldering, yet kept the high reliability of solder connections. The plug-in board and IC socket have fallen into disfavor among many manufacturers because oxidation of the contact pins causes intermittent connections and excess down time.

Figure 5–8 demonstrates the wire-wrapped connection. It consists of a wrap of several turns of wire around a terminal post. The connections made are highly reliable when made properly, and normally require no solder. Wire wrapping is also not destructive of components. Wraps can be easily unwrapped, and sockets easily reused. Though I

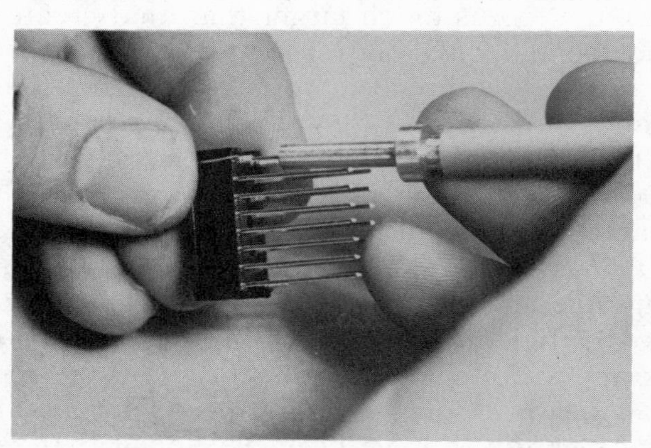

A

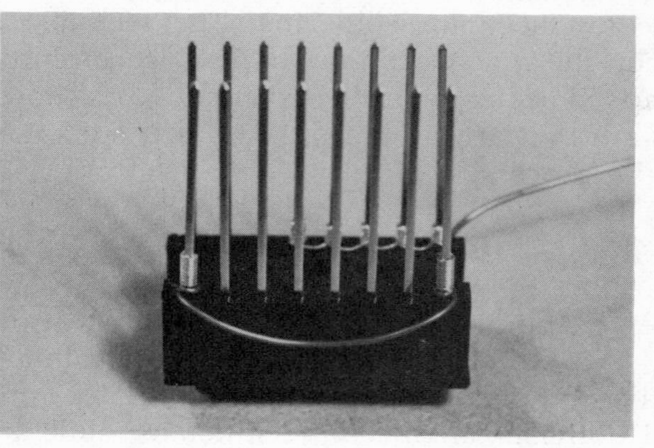

B

C

Figure 5–8
A. Using the Wire-Wrap Tool—Standard Wrap
B. Modified Wrap
C. Daisy Chain Wrap

have reused sockets many times, some technicians recommend that you solder any terminal that has previously been used. When you are prototyping, you will no longer have to throw away a bunch of components that were too much trouble to unsolder.

Wire wrapping works because the square-cornered post actually digs into the wire, making an airtight, oxidation-free connection. The posts are mounted in a special type of circuit board, known as a "perf board" or on a special wire-wrap terminal block. The perf board contains a grid of holes, consistent in diameter and spacing. Standard-size wire-wrap sockets and posts fit into the board, components on one side, and wiring on the other.

There are several basic wrap techniques, including the standard wrap, the modified wrap, and the daisy chain. The tools that make these connections are shown in Figure 5–9. Industrial tools are powered, and a pull of the trigger slits, wraps, and tightens the wire on the connection in an instant. Hand-wrap tools are more appropriate for the experimenter. They are inexpensive and reliable, usually costing less than fifteen dollars.

Using a manual wire-wrap tool is easy, and instructions

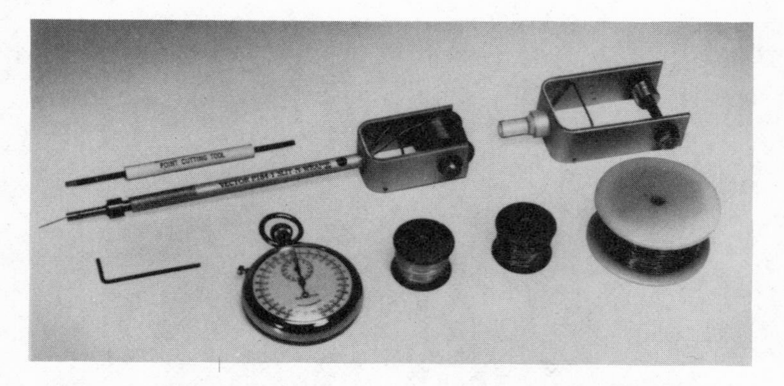

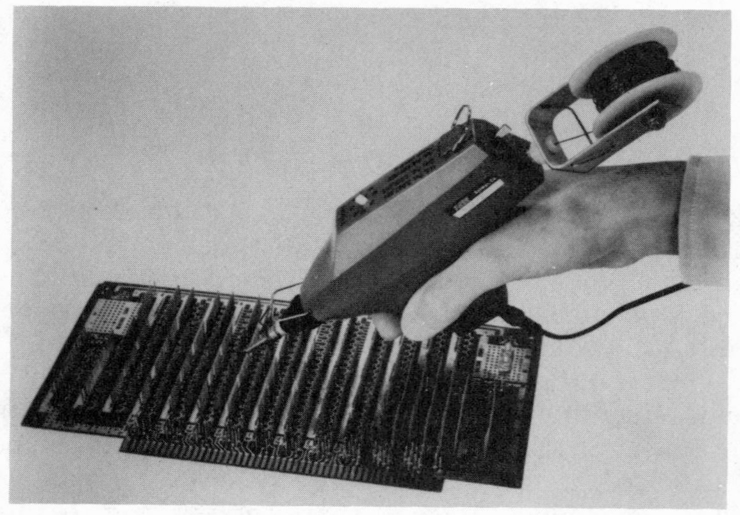

Photos courtesy of Vector Electronic Co.

Figure 5–9
A Manual Wire-Wrap Kit and an Electric Wire-Wrap Tool in Operation

are usually included with the tool. All tools are capable of the standard wrap, and most tools can perform the modified wrap. As you can see from the illustration, the modified wrap actually wraps two or three turns of insulated wire at the bottom of the post. This is done to strain relief the wire, protecting it from breakage if it should be flexed. The modified wrap is recommended in most cases, though it is not absolutely necessary.

Wire for wire wrapping is usually available in several sizes, but the most commonly available is 30 gauge. This wire is extremely small compared to most wiring in soldered projects. A little extra care is required, as these wires can be easily broken. Normally, wire is available in 50-foot or 100-foot rolls, and must be cut and stripped to length. A convenience for the experimenter is precut and stripped wire. This wire is packaged in kits of 1-inch to 8-inch wires, measured and stripped, ready to be installed in the wire-wrap tool.

Wire wrapping must be done on the square corner wire-wrap posts, and is best used when most components are integrated circuits. Resistors, capacitors, and other components with round leads must be soldered to the wire, or to wire-wrap posts. Remember, the major reason for wire-wrap reliability is that the post corners notch the wire. As a result, many projects that do not contain large amounts of ICs might be easiest to solder. An alternative technique is to wire wrap parts of the circuit, soldering those components that cannot be wrapped. Many of the projects that I build are constructed in this manner. I will probably never go back to making PC boards for those one-of-a-kind projects.

Wire-wrap tools can be purchased using the same rules you would use for purchasing any quality tool. The manual wrap tool should be nicely shaped and should operate easily in the hand. Most tools require prestripping the wire and inserting the prepared wire end into the hole at the tip of the wire-wrap tool. Placing the terminal on the post, and rotating in a clockwise direction will securely wrap the connection. To unwrap a connection, simply place the tool over the post that contains the wire to be removed, and rotate counterclockwise. What could be simpler?

The electric wire-wrap tool is constructed somewhat like a portable electric drill, but is a much lighter-duty tool. To operate an electric tool, just insert the prestripped wire into the bit, place the bit over the post, and pull the trigger on the tool. The motor will turn on and properly wrap the connection.

When selecting a tool, use those guidelines mentioned in the previous chapter on tools. The wire-wrap tool, like the soldering iron, is used to construct a finished circuit, capable of long-term operation. If you are just testing, or building circuits to see how they operate, or checking the operation of various components, you won't want to solder them into a circuit. Wire wrapping is somewhat better, but it is still a lot of work to lay out and wire the circuit. The experimenter's breadboard, or "protoboard" is the tool to use for any temporary wiring project. In the next section, we will look at important considerations in the use and selection of a protoboard.

Protoboards and Other Breadboard Techniques

In the early days of electronics, experimenters would connect circuits to old boards. They would pound nails into these boards, and solder components and connecting wires to them. These "breadboards" were so named because of their similarity to the old wooden boards used to hold bread, just out of the oven.

The experimenter would build a circuit in this fashion, and after it was completed, would test and troubleshoot it. After being assured of its proper operation, the experimenter would remove the components and wiring from the breadboard and install them in a chassis, or other enclosure, as a finished project. The value of this technique is still well recognized. The technology has changed tremendously, though, and the advent of the IC and solid-state electronics has given birth to a new type of breadboard. This prototyping board, or "protoboard" has rows of holes in a plastic surface. Underneath the holes are rows of connectors that allow one row of holes to be connected to another. Spacing is such that standard ICs and other components are easily inserted into the boards, then wires are inserted into the appropriate locations to complete the circuit.

Figure 5–10 is a photo of the protoboard, with a simple circuit wired on it. As you can see, the circuit needs no soldering, components do not need their leads cut, and all parts can be easily interconnected for testing. To use a protoboard properly, you must know exactly how it is wired internally. The diagram in Figure 5–11 illustrates how one

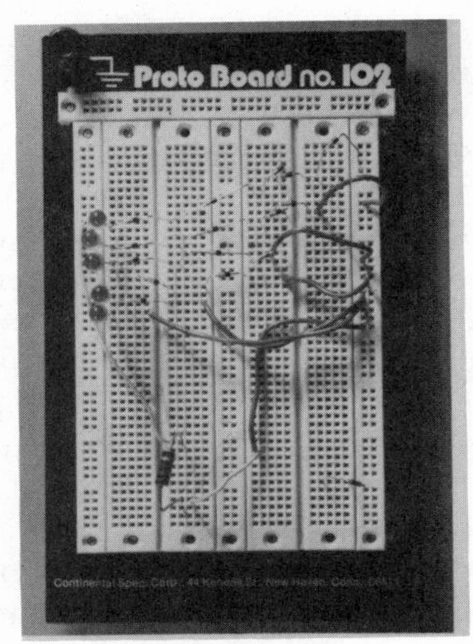

Figure 5–10
Prototype Circuit Development Board

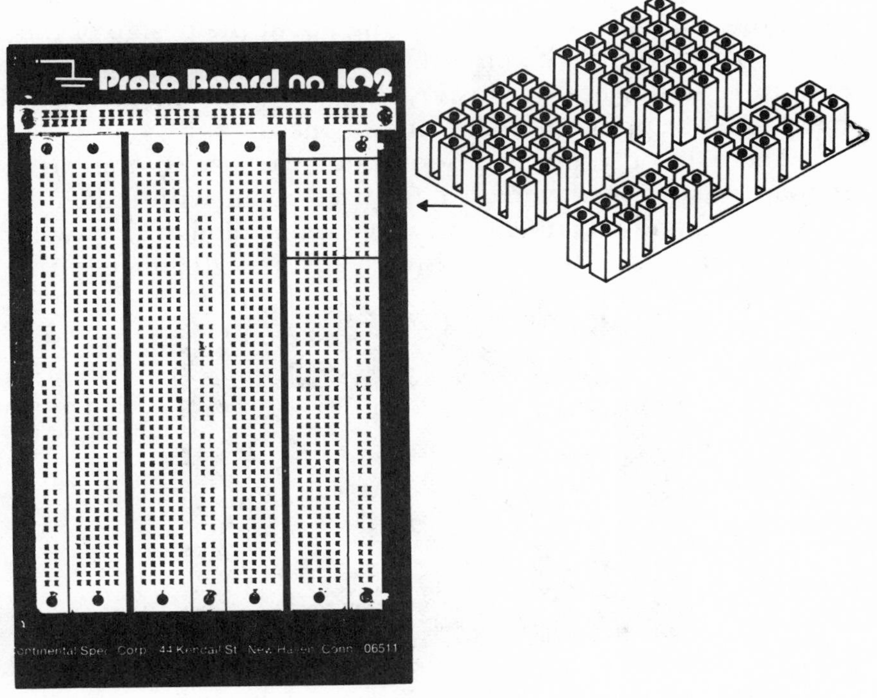

Figure 5–11
Diagram of Proto Connector Board

protoboard is wired. The two standard connector strips have two methods of internal connection. One type is for circuit wiring, and is divided into two banks of five connectors per row. Each row is connected across; columns are not connected. By inserting an IC into the center of the bank (straddling the gap in the center), you may independently hook to any pin on the IC. The small strips that contain only two columns and two rows are busbars. These are used for power distribution, and are connected, unlike the larger strips, along the columns. Typically, one line is used for positive voltage, and the other for negative. Most boards have two or more bus strips so that multiple-output voltage can be connected to the circuit.

A variation on the protoboard is the circuit development system. Figure 5–12 is a typical circuit commercially available. The heart of the system is a protoboard, but the device also contains several power supplies, and possibly, signal generating circuits. The experimenter can develop operating circuits without worrying about the design of the power supply until the circuits actually work. Though the system is somewhat expensive, the avid experimenter will have lots of applications for one of these.

There is only one caution about the use of the protoboard. As an instructor, I have used them with students and have found a couple of weaknesses. Although they will accept standard 1/2-watt resistor leads and 20-gauge wire,

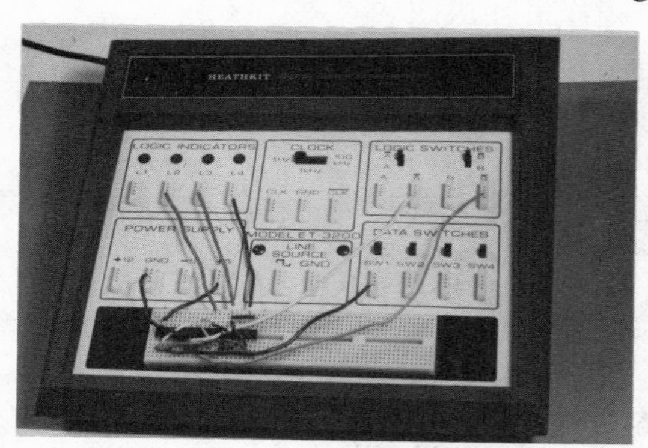

Figure 5–12
Prototype Circuit Development System with Power Supplies

they are really designed for a smaller gauge wire. A 1/4-watt resistor and 22-gauge wire are recommended for maximum life of the connector strip. Large component leads stress the contacts, and if they are used repeatedly for larger wires, they will not connect properly when being used for integrated circuits and other smaller components with small lead sizes. If you need to use a component with heavy-duty leads, solder 22-gauge wires to the component leads, and use these to connect to the circuit. An even better idea is to buy a set of test leads, which contain miniature alligator clips on each end. Keep a few around for general purpose, but take two cables and cut them exactly in two. Strip the cut ends and solder them individually to the end of a 1/4-watt resistor. Cut the resistor away and you will have small wires that can be inserted into the protoboard without damaging it. Keep these for use when large components must be installed. Just clip the desired component into the alligator clips, and set it beside the perfboard.

If you plan on eventually working with high frequency circuits specify that a metal base or "ground plane" be underneath the protoboard. Most boards have this available, however some development systems do not. High frequency circuits can be unstable and there can be some undesirable effects. The ground plane helps to keep the circuit stable.

If you decide to purchase a development system, there are two basic types available. Be sure to get the one that meets your needs most of the time. If you plan on working with amplifiers and other linear circuits, the development system should contain at least one infinitely variable supply that goes to at least 25 volts. A digital circuit experimenter usually will require only a 5-volt, and a positive and negative 12-volt, three-output supply. Current capacities vary somewhat; however, most available systems have enough current reserve to supply the typical circuits wired onto the protoboard. Be careful with current capacity, and don't use the supplies to power external devices unless you are sure a device will not exceed the capacity of the supply.

If you plan on doing a lot of experimenting with wiring of newly designed circuits, you can easily justify the purchase of a protoboard, and will probably put a circuit development system to good use. The singular advantage of circuit testing and checkout using a protoboard is its time-saving technique for wiring and troubleshooting experimental circuits.

You are just about ready to start building your own complex circuits, and the next chapter will describe the effects of alternating current in a circuit. When you have an understanding of the concepts of AC circuits, you will be ready to take on the power supply (and build one if you like). Understanding and application of the transistor will follow.

Understanding AC Circuits

6

In Chapter 1, we looked at some of the differences between AC and DC. This chapter will look in more detail at those differences. The effect of coils and capacitors on AC circuits is especially important, and is covered here. AC circuits that use coils and capacitors will be affected in ways not applicable to DC circuits, and their rules and effects will be explained. The concepts of reactance and impedance are the main topics of this chapter. If you have ever bought a set of speakers, you are probably familiar with the term impedance, even if you haven't understood exactly what it means. The term "reactance" is not used quite as often, and you may not have come across it before. It would be a good idea to review the section on the differences between AC and DC in Chapter 1 before we go on.

Why We Use AC

The reason for using AC is the ease of transferring power from the source to the home. This, however, is only one small part of the picture. The major use of AC signals is the production of audio frequency and radio frequency signals. These signals are electrical representations of the output of devices such as tape recorders, radio receivers, television receivers, transmitters, and all manner of communications circuits. The frequency spectrum is represented in Figure 6–1. The spectrum chart also lists some of the commonly known services at each part of the spectrum.

One of the major advantages of AC, especially at higher frequencies, is the ability to change from electrical signals to electromagnetic radiation. This ominous sounding phrase simply means that it is easily possible to send these frequencies through the air to be received by appropriate receivers. Before looking at this in more detail, let's look at the characteristics of AC.

There are two basic characteristics of all AC signals. These are amplitude and frequency. Look at Figure 6–2 for a pictorial of these parameters. Amplitude is essentially what is measured when we hook up a voltmeter to the circuit. In other words, the circuit pictured might have an amplitude of 10 volts. This voltage can be measured from the peak of the upper part of the wave to the peak of the lower wave. This form of measurement is called "peak-to-peak." A "peak" voltage reading would be from the top of the wave to the exact center. Peak voltage will be exactly half of the total peak-to-peak voltage. The negative peak value would be from the center of the waveform to the bottom. A 10-volt peak-to-peak voltage would be +5 volts to the positive peak, and −5 volts to the negative peak. (See p. 156.)

Many waveforms vary from zero to positive, and back to zero, then go negative and return to zero. It is possible to have an AC voltage vary around a voltage other than zero. The center voltage will be a certain level, for example, 20 volts, and the AC will vary, in this case, from +25 volts to +15 volts, still a 10-volt swing, though the swing is not from positive to negative. Amplifiers use this characteristic to "bias," or provide DC operating voltages to the transistors, while they amplify the AC signals that are sent to the devices on the bias lines. More on this later.

Another important point you should be aware of is the fact that the voltmeter does not measure the voltage peaks in a circuit. The meter cannot respond quickly enough to catch the peaks of voltage. It is calibrated to read an average

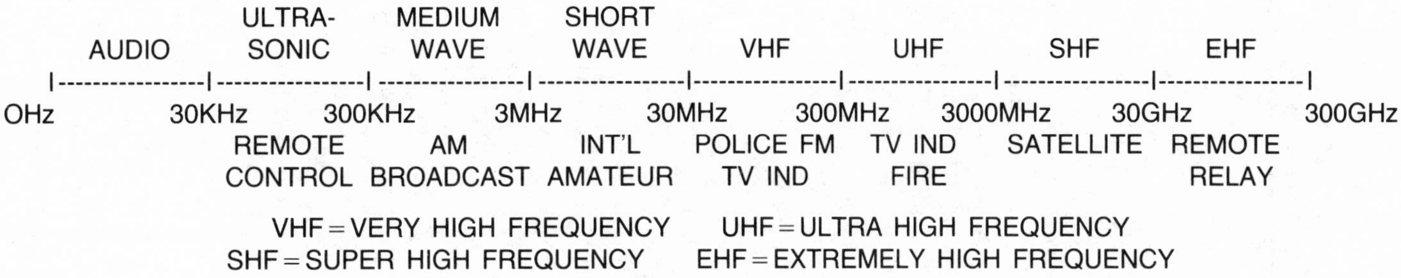

AUDIO	ULTRA-SONIC	MEDIUM WAVE	SHORT WAVE	VHF	UHF	SHF	EHF
0Hz – 30KHz	30KHz – 300KHz	300KHz – 3MHz	3MHz – 30MHz	30MHz – 300MHz	300MHz – 3000MHz	3000MHz – 30GHz	30GHz – 300GHz
	REMOTE CONTROL	AM BROADCAST	INT'L AMATEUR	POLICE FM TV IND	TV IND FIRE	SATELLITE	REMOTE RELAY

VHF = VERY HIGH FREQUENCY   UHF = ULTRA HIGH FREQUENCY
SHF = SUPER HIGH FREQUENCY   EHF = EXTREMELY HIGH FREQUENCY

Figure 6–1
Frequency Spectrum Chart

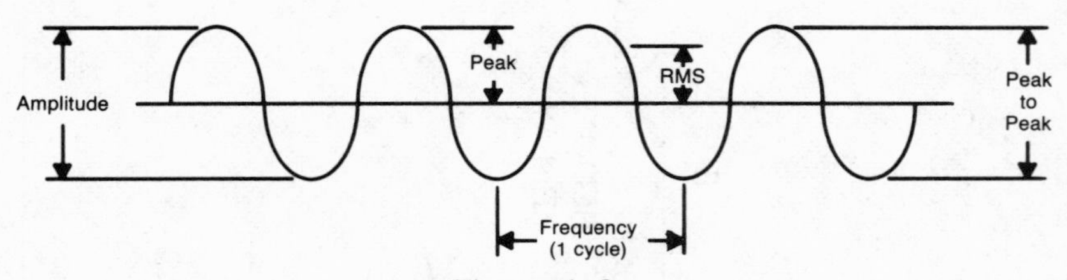

Figure 6–2
Diagram of Sinewave

amount of voltage, which really represents the actual amount of equivalent DC voltage that is available. Remember that the actual work being done in a circuit is expressed in the formula: P = E × I. Wattage is equal to voltage times current. If you were to calculate the power using peak values, you would find them to be inaccurate. Let's see why. If you use the value of 10 volts in your circuit, and you have a 10-ohm resistor as a load, Ohm's law states that 1 amp of current will flow. That means that 10 watts would be dissipated in the circuit. If you have an AC ammeter, though, you will find that the current is actually far less than 1 amp. This is because the circuit is delivering only 5 volts, maximum, to the circuit. This 5-volt measurement is not constant either, and the actual work being done in the circuit is really much less than the calculations would have you believe.

To get around this problem, engineers have calculated the actual work being done in a circuit, and expressed this as the effective, or RMS, voltage. All voltmeters express sine wave values as RMS values. There is a mathematical relationship between the effective and peak values of a true sine wave. A 5-volt peak sine wave will actually produce the power output of a 3.5-volt DC circuit. The formula for converting peak voltage to RMS voltage is PEAK × 0.707. Conversely, RMS /0.707 will calculate the peak voltage of an AC waveform. If you work with an oscilloscope, you will have to keep these numbers in mind. The scope is calibrated in peak values, while the meter, as was stated, reads the RMS value. You must provide the conversion.

The other parameter mentioned in Figure 6–2 is frequency. The frequency is the amount of time that it takes for one cycle to be completed. For example, an AC waveform that changes from 0 to +5, back to 0, and down to −5, then to 0 again, is defined as one cycle. If 1000 cycles occur in 1 second, the frequency of the waveform is considered to be 1000 cycles per second, 1000 hertz, or 1 kilohertz. If you know the frequency, you can calculate the actual time it takes to complete one cycle. The formula to calculate the time period is: T (in seconds) = 1 / hertz. In our example, the period is 0.001 seconds, or 1 millisecond. Looking at the frequency chart, you have probably realized that this is a relatively slow frequency. As was stated earlier, frequencies of several thousand megahertz are used daily in radio and television communication.

One other important parameter of an AC waveform is its phase. Phase is time related, and refers to the actual voltage or current at any given time. If you were to fold the sine wave over upon itself, it would be drawn as a circle. Refer to Figure 6–3 for the following description. Experimenters think of the start of the cycle as zero degrees. On a typical sine wave, this would be the zero volt point. At 90 degrees, the voltage would be at maximum. At 180 degrees, the waveform has progressed from its maximum positive voltage to 0 again. The most negative part of the waveform occurs at 270 degrees, and the voltage returns to 0 to complete the 360-degree cycle. Remember the following two important points—the 180-degree point is reached only during the positive-to-negative 0 transition, and 360 degrees and 0 degrees occur during the negative-to-positive transition. Also, the term "360 degrees" is used when you wish to

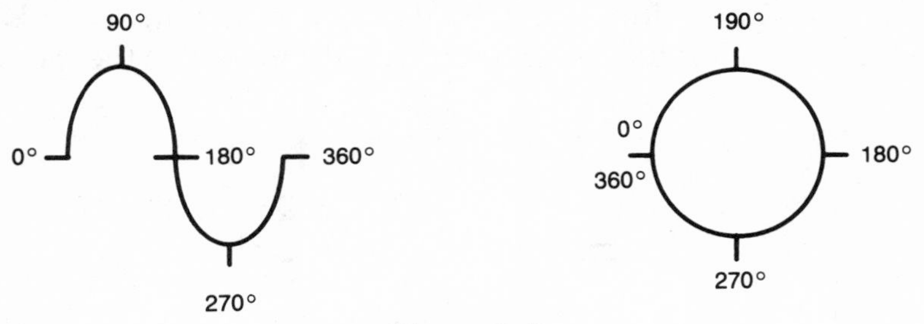

Figure 6–3
Diagram of Phase Relationships

describe the end of the previous cycle, while the term "0 degrees" is used to describe the start of a new cycle. Actually, they are the same points in a waveform.

Now that we have looked in more detail at the actual AC waveform, let's see how various components are affected, when they have AC applied to them. A new definition is about to be presented for a property that is quite similar in many ways to resistance. Reactance is the term; let's see exactly what it is.

What Reactance Is

It was just implied that reactance is somehow related to resistance. That is correct, up to a point. One of the easiest ways to think of reactance is as resistance to AC. Reactance is actually the opposition to changing voltages or currents. The key word here is *changing*. DC circuits have no reactance, with the exception of two special times.

Let's examine the circuit of Figure 6–4. It is a simple DC circuit, with a coil instead of a resistor. If you turn on the circuit via switch SW1, the DC voltage supplied by the battery will be applied to the coil. The voltage across the coil will try to rise immediately to the value of the battery. However, a strange phenomenon occurs. A countervoltage is developed in the coil. The countervoltage, dubbed "counter EMF" actually opposes the voltage being applied to the coil, and slows its rise to maximum across the coil.

The amount of time the DC across the coil takes to rise to the same level as the battery voltage can be easily calculated with one simple formula: $T = (L / R) \times 5$. T in seconds is equal to the value of inductance divided by the actual

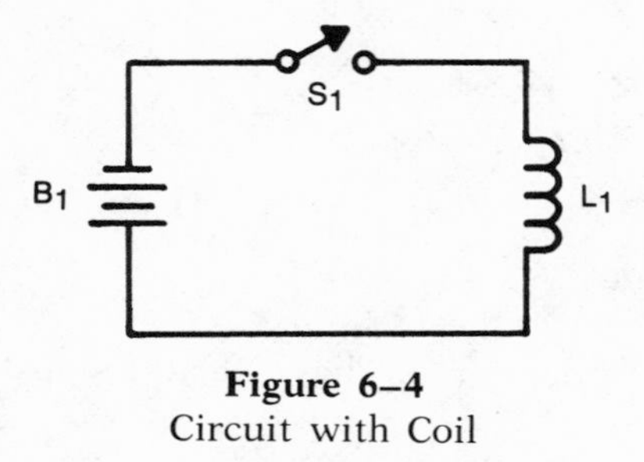

Figure 6–4
Circuit with Coil

resistance of the coil (including any external circuit resistance) times 5. The change in voltage is actually delayed by the reactance of the coil, opposing current flow in the circuit. Reactance caused by a coil is known as "inductive reactance," and is defined as the opposition to change in current. Inductive reactance is developed when the sudden rush of current through the coil creates an intense magnetic field. This field self-induces the opposing voltage that slows the rise of current. During the time that the DC is unchanging, there is no opposition to changes in current, and therefore inductive reactance is not present. The only opposition to current flow, once the voltage reaches full value, is from the resistance of the coil. The only other time that DC is affected by reactance is (you guessed it) when the circuit is shut off. The sudden change from maximum voltage to minimum again develops a counter EMF that opposes the changing current.

As you have probably already figured out, the operation of a coil in an AC circuit is quite complex. Every time the voltage rises above zero, inductive reactance opposes the change in current that is trying to take place. On the other hand, before the coil stabilizes, and lets current flow, the AC voltage is decreasing again. Again the counter EMF in the coil develops, this time aiding the flow of current. What this actually means is that the coil will always have two AC signals across it—the actual applied voltage, and up to 90 degrees later, the current waveform.

It is beyond the scope of this book to go into great detail about the effects of reactance on AC circuits, except where they affect the experimenter. The basic concept that you should remember is that as frequency increases, the reactance of a coil increases. This relationship is direct, and easily calculated. The formula is: Xl = 2 pi × F × L. Inductive reactance (Xl) is equal to 2 pi times the frequency (in hertz) of the AC waveform, times the inductance (in henries) of the coil. Coils, then, must be carefully selected so that they have the proper amount of reactance at the frequency at which they are to be used.

There is another type of reactance that the experimenter works with. Capacitive reactance is defined as the opposition to changes in voltage in a circuit. AC circuits containing capacitors are affected by this parameter. Interestingly enough, the effects of Xc are exactly the opposite of those of inductive reactance. Unlike Xl, Xc actually decreases with increasing frequency. In other words, the higher a frequency, the more easily it will travel across a capacitor. The formula that will solve for capacitive reactance is:

$$Xc = \frac{1}{2\ pi \times f \times c}$$

or, capacitive reactance is equal to the reciprocal of 2 pi times frequency (in hertz) times capacitance (in farads).

Circuits that contain capacitors also have two waveforms, one representing the current in the circuit, and one, 90 degrees later, representing the voltage. This is the opposite effect to that noted regarding inductors. An old saying expressing the changing waveforms is "ELI the ICE man." The E and I represent voltage and current respectively, and the L and C represent inductive and capacitive reactance. The sentence states the following relationships: Voltage leads current in an inductive circuit, while current leads voltage in a capacitive circuit.

We've covered a lot of territory in the last few paragraphs, and we have really only scratched the surface of one of the most complex topics in the area of electronics. If you desire further study in this area, refer to a college-level text. The experimenter must keep in mind that reactance is a form of AC resistance. Its effects are seen in working with AC circuits, and the individual components that cause reactance are inductors and capacitors. The next section looks at how reactance is used in a circuit to affect the operation of those circuits, and the components in them. We will also look at ways that we actually use reactance to assist in certain operations.

How Reactance Affects Components in an AC Circuit

As we have seen, reactance relates to capacitors and inductors, and is dependent upon the frequency. Let's look at some of the uses we have for reactance. This section will cover, specifically, inductors and capacitors that are connected in AC circuits.

If you install an inductor in series with an AC source, and then connect a capacitor to ground, as in Figure 6–5A, frequencies above a certain point will be held back by the coil. Those high frequencies that get through the coil will be shunted to ground via the capacitor. This is the basic design of a low-pass filter. This circuit is used whenever the experimenter wishes to block high frequencies from reaching a component or circuit.

Connecting a circuit such as that in 6–5B will result in low frequencies being blocked by the reactance of the capacitor. Any that make it past the capacitor are also shunted to ground through the coil. The high-pass filter is used when it is desired to limit the flow of low-frequency AC in a circuit. A practical example of a high-pass circuit is the crossover network in a speaker system that allows high frequency AC to reach the tweeter, while blocking the lower frequencies.

There is an interesting characteristic in connecting coils and capacitors. We can start by making a single frequency "trap." A trap is actually a filter circuit that allows the experimenter to remove an undesired frequency from a circuit. One example of a trap is the filter that keeps the audio portion of a TV broadcast from interfering with the video. You can partially detune the trap in your TV set by misadjusting the fine tuning. If you fine tune in one direction, you will begin to see an interference pattern on the screen that is caused by audio affecting the picture.

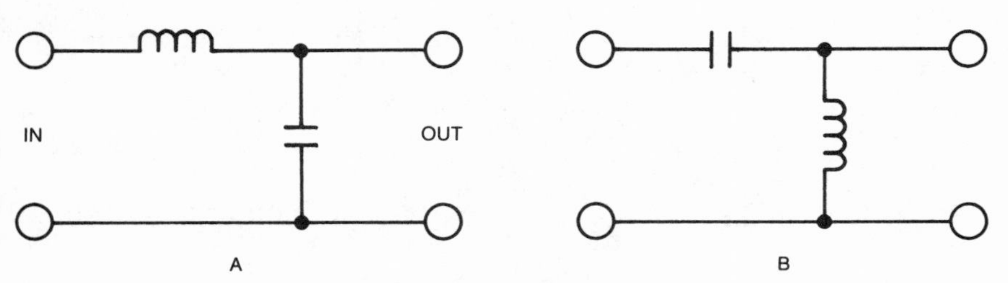

Figure 6–5
Diagrams of Low and High Pass Filters

Another example of a trap is the filter included in all CB radios that keeps the radio from interfering with TV Channel 2. It is tuned at the factory for best rejection at the frequencies covered by Channel 2.

Figure 6–6 shows how simply a trap can be built. The conductors carrying AC into the trap circuit can contain one or several frequencies. The reactance of the capacitor will block out low frequencies, not allowing them through to the coil. The reactance of the inductor will block the high frequencies, not allowing them to travel to circuit ground. There is, however, one group of frequencies that is in the right range, that will flow through both the capacitor and coil, and be trapped, or shunted to ground. The response curve in the figure displays the output of the trap. As you can see, frequencies below the "resonant frequency" that the trap is tuned to, will have a high voltage at the output. As the frequency increases, the reactance of the trap decreases, allowing the undesired frequency to flow through to ground. Frequencies above Fr (frequency of resonance) will also be allowed through the trap.

As you can see from the curve, there is actually a range of frequencies that are trapped out. Experimenters and others have determined that a frequency that is at least 0.707 times the maximum, or peak, voltage is not being trapped by the circuit. The range between the lower and upper frequency, where the effects of the trap are negligible, is called the bandwidth. An example might make it just a

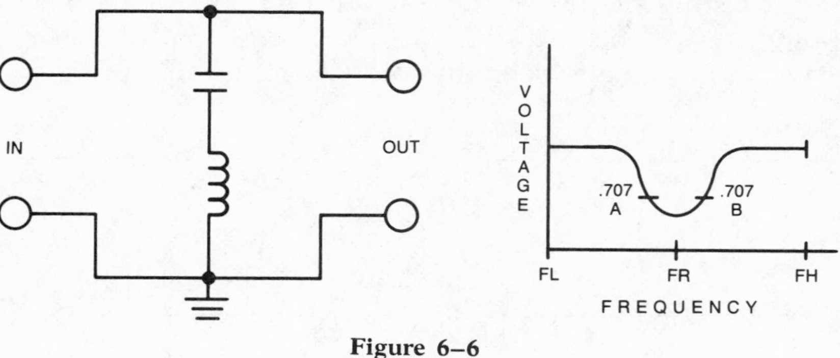

Figure 6–6
Trap Circuit

little easier to understand. Let's assume that the resonant frequency of this circuit is 1000 Hz., point A is 900 Hz., and point B is 1100 Hz. The bandwidth of this trap is 200 Hz.

Coils and capacitors have a rating that is called Q, for quality. This rating is frequency, as well as component, dependent and is a comparison, between the actual resistance and reactance at a given frequency, of the component. High-Q circuits have a very narrow bandwidth, affecting a small range of frequencies, while low-Q circuits are broad band, and allow a large range of frequencies into the trap. As you have probably already guessed, there are applications for both types of circuits.

Figure 6–7 represents a parallel tuned circuit. This type of resonant circuit is known as a band pass filter. It is opposite in operation to the trap. In other words, only the desired bandwidth of frequencies is allowed through the filter to the output. Frequencies below Fr pass easily through the inductor to ground, while frequencies above Fr travel to ground through the capacitor.

The concept of reactance offers many possibilities. These properties of coils and capacitors provide all the capabilities for communications electronics as we know it today. Designers, choosing the right combinations of "tuned circuits," circuits containing coils and capacitors, can accept or reject any frequency desired. This principle is used in channel selection for radio and television reception, tone control in stereo and audio systems, and in many other applications.

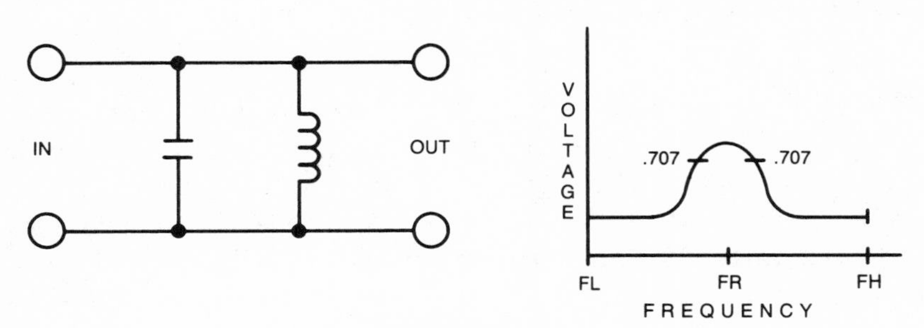

Figure 6–7
Parallel Tuned Circuit

We have seen how to put reactance to good use in AC circuits, and will now look at impedance, to see just exactly what it is and how experimenters use it.

What Impedance Is

After looking at the combination of resistance and reactance, you have probably already assumed that there must be another quantity related to opposition to current flow. Since resistance is a property of DC and AC circuits, while reactance is a property of AC circuits only, how can we express the total opposition to current flow in an AC circuit? You guessed it—impedance is total opposition to current flow in an AC circuit. Like resistance and reactance, impedance is measured in ohms, and its symbol is Z. The total opposition in an AC circuit is always a combination of total resistance and reactance.

An AC circuit containing no coils or capacitors can be considered to have no reactance, therefore the impedance is equal to the resistance of the circuit. You can calculate total opposition to current flow using only the resistance. Things get a little more complex when reactance is involved. Remember, it was stated that circuits containing reactance have the voltage out of step with the current. In inductive circuits, current lags behind the voltage, while in capacitive circuits, voltage lags behind the current. In either case, we cannot simply add up the resistance and reactance to calculate total opposition to current flow. In series circuits, total opposition can be calculated by the formula

$$Z = \sqrt{R^2 + X^2}$$

Calculating impedance in parallel circuits can be a complex process, however there is a shortcut that will provide you with the answer. Before you use the formula above, first divide each reactance into one. Then take each of these reciprocals and use those values as R and X in the formula. After calculating for Z, the actual value of Z is obtained by dividing the result into one. Let's try an example. A parallel circuit has a resistance of 5 ohms, and a total reactance of 10 ohms. Before using the formula; divide 5 into 1 and 10 into 1,

$$1/5 = 0.2 \text{ and } 1/10 = 0.1$$

Enter these values into the formula.

$$Z = \sqrt{0.2^2 + 0.1^2}$$

Calculating these values makes Z = 0.224 (approximately). The last step is to calculate total Z by dividing the result into 1.

$$Z = 1 / 0.224 = 4.46$$

Total impedance of this circuit is 4.46 ohms.

Total reactance in a circuit containing both inductors and capacitors can be figured by subtracting.

$$Xt = Xl - Xc$$

The total impedance in a circuit can be easily transferred into those Ohm's law problems for DC with direct substitution. In other words,

$$E = I \times Z.$$

Impedance will substitute directly for resistance.

The only exception to this general rule occurs in working with AC power. Since currents and voltages are not in phase with each other, power calculations using impedance yield a value called apparent power. Apparent power is measured in volt-amps, and will not express the actual amount of work being done in the circuit. Unless you want to go into a lot of trigonometry to calculate the true power, about the only method you have to determine actual power being delivered into the circuit is to break it down into components. Measure the currents and voltages across the resistive components. Calculate power from these values.

Dealing with impedance is relatively easy; unfortunately, it is a difficult quantity to measure. To measure impedance properly, the equipment needed is an impedance bridge. These are not readily available, and fortunately, they are seldom needed. The next section tells you why impedance is important to the experimenter, and how he or she can use it to advantage.

How Impedance Affects Components in AC Circuits

Though the concept of impedance can be complicated at the engineering and design level, the experimenter has to work with only a couple of important concepts. If you have ever bought a stereo component system, you have certainly heard the term. Usually it is used when referring to speakers. As an example, the typical stereo requires 4-ohm to 16-ohm speakers, with 8 ohms being the most common. When you purchase speakers for your stereo, you must select the correct output impedance.

In order that maximum power transfer may take place between components, you must match impedances. In other words, the transfer of energy from one circuit to another is best accomplished when the output of one circuit matches the input of the next. Transistor and tube circuits have relatively stable output impedances, which vary depending upon the exact hookup of the device. Figure 6–8 contains a block diagram of a typical audio matching circuit. The output stage contains either a tube or transistor which must be matched to a speaker. The primary of a transformer is matched to the output stage, while the transformer secondary is matched to the speaker. As you can see, this is a relatively simple and easy way to match impedances. Newer transistor equipment often tries to match impedances in other ways, because of the relatively high cost of transformers. Most of the circuits designed for this book do not use transformers for impedance matching. The chapter on transistors develops this topic in more detail.

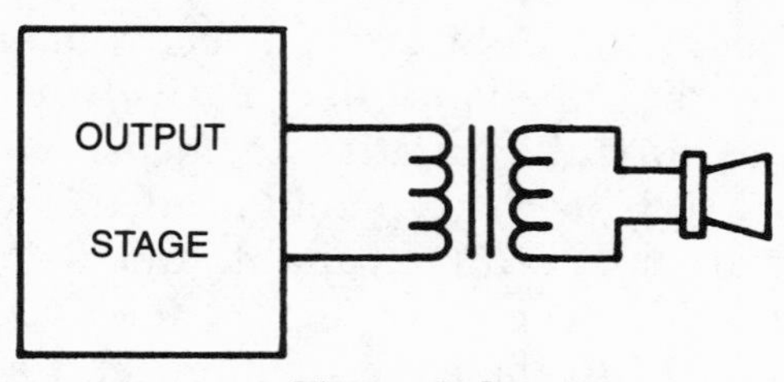

Figure 6–8
Block Diagram of Audio Circuit

Audio devices are generally classed as high or low impedance devices, and the experimenter must match them up when buying accessories. As impedance gets higher, it becomes somewhat less critical that there be an exact match; however the closer you can come, the better.

Another device that requires proper impedance matching is the radio transmitter. If you have had any experience with a CB radio, you have run into this, when hooking up the antenna. The CB radio has a 50-ohm to 75-ohm output circuit, and the antenna must be tuned properly to allow the CB to deliver all of its available power to the antenna. The tool used to measure antenna matching is called an SWR bridge. SWR is an acronym for the term "standing wave ratio," and is the ratio of maximum current to minimum current along a transmission line. Effectively, it is the measure of the mismatch between the antenna load and transmission line. When the SWR is improperly matched, transmitter power is used to heat up the transmission line and output circuits of the transmitter. We would much rather use the transmitter power to radiate as much signal as possible from the antenna and into the air.

All circuits that have high reactance or high resistance, or both, will have a high impedance. We can see that from the formula in the last section. The actual value of the impedance depends upon the circuit involved. The series tuned circuit discussed previously has a low impedance at the resonant frequency. This means that the frequency the circuit is tuned to will develop relatively high currents, compared to frequencies that are not resonant. Remember, as with resistance, impedance and current are inversely proportional.

The parallel tuned resonant circuit has a high impedance to the frequency of resonance, therefore low currents flow at resonant frequency. Frequencies far from resonance will draw high currents through either the inductor or the capacitor, depending upon whether the frequency is below or above the frequency of resonance.

Many times the experimenter runs across the effects of poor impedance match, without realizing it. In some cases, there is little if any serious problem, especially if the mismatch is small. There are times when impedance matching is critical. One of the cases, mentioned earlier, is the CB radio. A CB radio or any transmitter can actually be damaged by not having proper impedance matching. The high line currents on a mismatched antenna line can actually cause the output transistor to overheat and burn out if care is not utilized. Many transmitters, as a result, have special circuits that lower the power levels to the transmitter output devices. These circuits measure the output reflected power level, and if it exceeds a certain preset value, the protection circuitry shuts down the output current of the transmitter, protecting it from damage.

The concepts that have been covered in this chapter are complex, and engineers continue to study their effects on newly designed circuits. The hobbyist and experimenter must be aware of the terminology, and have a basic understanding of reactance and impedance. All circuits designed must take these factors into account, as must any devices you build.

We have covered a lot of material in the last few chapters, and now it is time to put our theory to work. The next chapter discusses the power supply, and how it is designed and constructed. By applying the rules you have learned in this and the last few chapters, with the information in Chapter 7, you should be able to design and build a power supply that will meet your own individual requirements.

How Power Supplies Work

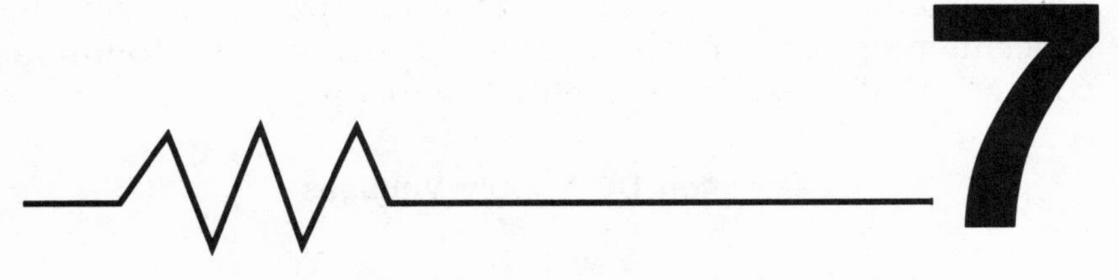

How Experimenters Use Power Supplies

The electronics hobbyist is often working with power supplies and the circuitry involved in them. Fortunately, the power supply can be one of the simplest, most reliable, and easiest-to-understand circuits found in electronic equipment. Though there are complex power supplies, they are rare in experimenters' circuits, so most emphasis will be placed upon understanding the components in the basic power supply. The functions and stages in the supply, and design considerations you can use to help construct working supplies that meet your experimental needs, will also be emphasized.

The basic purpose of the power supply in any piece of electronic equipment is to supply each operating stage or circuit with the necessary voltages and currents to allow the equipment to continue to operate properly. With this in mind, let's look at the basic functions of the power supplies.

Almost all active electronic devices, transistors, tubes, etc., require some value of DC voltage or voltages to operate, yet the voltage most homes and businesses are supplied is 120 volts AC, which is a relatively high voltage for small solid-state components. From these last statements, you have probably already determined the two functions of power supplies. They are:

Change alternating current to direct current.
Provide proper voltage levels for the various stages.

The experimenter will have to use both functions whenever he designs a power supply. Most circuits that experimenters use will operate at certain standard voltage levels. You could design a circuit around a nonstandard voltage, and many companies do just that; however, they have easier access to components that meet their requirements than the experimenter. See Table 7–1 for a list of the common voltages that most experimenters will use.

Common DC Supply Voltages

+5 VOLTS
+12 VOLTS
−12 VOLTS
+13.8 VOLTS

Table 7–1
Standard Supply Voltages

Referring to Table 7–1, note that these standard voltages are used for various purposes. The 5-volt supply is used extensively in computer and digital logic circuits. If you plan to experiment in this area, a 5-volt supply is a necessity. The +12 volt supply is used for general purpose circuits. It is high enough so that current levels can be smaller, for a given power output; yet most available components will accept this voltage. The −12 volt supply is often used in conjunction with the +12 volt supply in a dual output mode. Many linear integrated circuits require a dual positive-negative voltage supply. The 13.8 volt supply is the standard value used in automotive electronics. The standard automotive battery is 13.8 volts and a bench power supply that substitutes for the battery is a handy piece of equipment for the experimenter.

Many experimenters have built or purchased adjustable voltage power supplies, so that they can adjust the output to whatever level they require. These supplies are more complex in many cases, and are used only during the design of equipment. Once the levels are chosen, the designer usually builds a fixed value supply to permanently power the circuit. Also, it is not uncommon to find power supplies that have more than one voltage output. How an experimenter designs a multiple-output supply will be covered in the last section of this chapter.

Figure 7–1 contains the block diagram of the basic structure of the power supply. The power supply shown is a 13.8-volt output, 120-volt input supply. Starting from the left, or input, of the power supply, the first block contains the power transformer. This stage functions in a step-down mode. As you recall, a step-down transformer takes a higher voltage and transforms it to a lower voltage. The rectifier block contains one or more diodes, and its function is to change the low voltage AC into pulsating DC. This will be explained further in the next section. The filter stage removes the pulsations and leaves a pure DC output, which

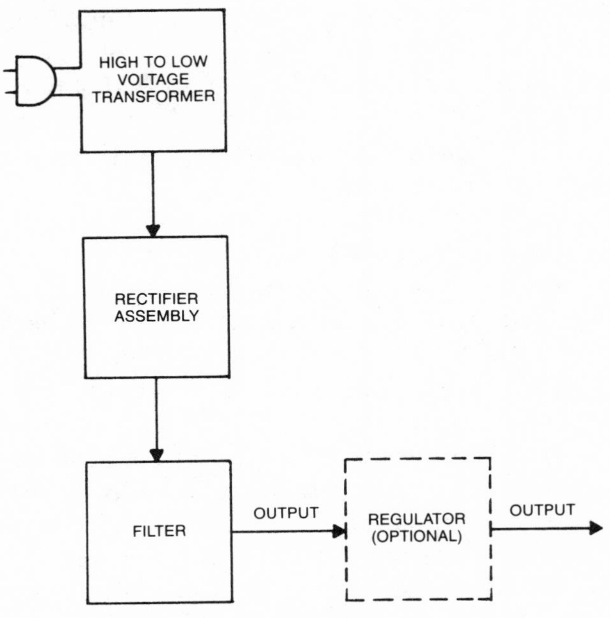

Figure 7–1
Block Diagram of Power Supply

can then go to the circuit being powered. Notice the block in dotted lines marked REGULATOR. This block is in some power supplies and its function is to provide a constant output voltage. The need for this block is dependent upon how critical the circuit is. If the circuit being powered needs a constant voltage, the designer must regulate the supply.

As we go through each block, we will explain the individual circuits, so that you will have a thorough understanding of the operation of each stage. The last section of this chapter is devoted to integrating the information into a format that will allow you to design your own power supply and put it into operation.

Starting with the first block, you will find the transformer. As you already know, the transformer can step up or step down the input voltage. Most power supplies that experimenters work with have 120-volt primary, however some transformers have a dual primary whose function is to operate with either 120-volt mains or 240-volt mains. See Figure 7–2A. The transformer (T1) provides one other basic function in a power supply. Because there is no direct electrical connection between the primary and the secondary windings, the transformer effectively isolates the circuit connected on the secondary side of the transformer from the AC line. As a result, the entire project, including the circuit being powered, is not directly connected to the AC line. This is a definite safety advantage.

The transformer circuit (Figure 7–2B) also contains some other useful components. These are the AC power switch (SW1) and an overload protection fuse (F1) to protect the input circuit from excess current flow in the supply.

Note that the power transformer is not always required; for example, if you need 120 volts DC, you could just directly connect the 120 volts AC to the rectifier circuit—using proper safety devices, of course. Many televisions and table radios use this method, as it results in lower production cost. One has to be careful, though, with this type of supply. As was stated earlier, one of the functions of a transformer is to isolate the circuits from the AC line. By omitting the transformer, you are also removing this important safety factor. When this is done, all circuits in the power supply, as

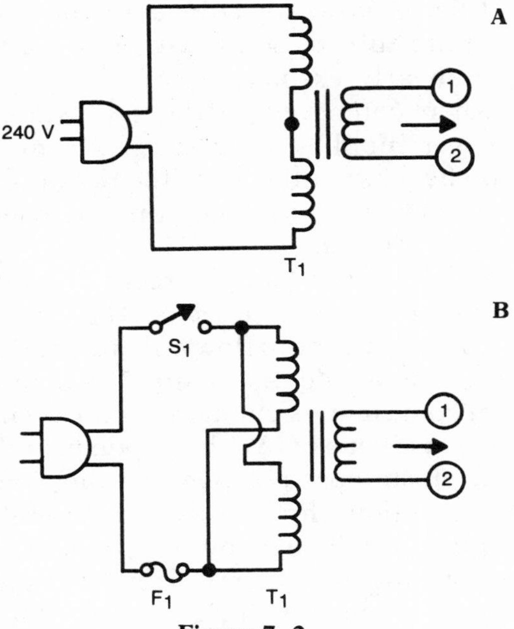

Figure 7–2
Transformer Stage Diagram

well as the circuit being powered, must be enclosed in a nonconductive cabinet, and any shafts, controls, or exposed parts must be protected by insulating material. When constructing or servicing power supplies of this kind, be sure to review, or be familiar with, the cautions given earlier in the section on electrical safety in Chapter 1.

How Rectifiers Work

Once the AC voltage leaves the secondary of the power transformer block, it travels to the rectifier block. The purpose of this block is to change the AC voltage into a pulsating DC. This will be described further as the output circuit is explained. You will learn about the three kinds of rectifier configurations, the half wave, the conventional full wave, and the full wave bridge. After you complete the chapter, you will be able to decide when each type is appropriate for your needs.

The rectifier is basically a diode. As you know, the diode could be vacuum tube or solid state, however the vacuum-tube diode is nearly extinct in new design. It is just too inefficient when compared to the solid-state diode. As a result, all power supplies discussed will contain solid-state diodes. If you are unfamiliar with the operation and testing of the diode, this would be a good time to review the section in Chapter 2 on diodes.

The simplest power supply rectifier circuit contains only a single diode. See Figure 7–3. The diode, as you will remember, allows current to flow only in one direction, from the cathode to the anode. As Figure 7–3 demonstrates, the AC waveform coming in is changing polarity and voltage 60 times per second (60 hertz). With a diode installed as shown, when the waveform goes positive, the diode conducts, and allows current to flow through load resistor RL. RL represents the equipment being powered, whether it is a dial light, or a solid-state amplifier. Current flows from negative to positive, and a positive voltage appears on the cathode of the diode. You may ask: If the cathode of a diode is considered negative, then why is there a positive voltage on the cathode? This is because the voltage is referenced, or measured compared to negative, or ground (the lower side of RL). Since the load resistor offers opposition to current, there is a voltage drop across it, which is actually what we are measuring at the diode. If you were to measure the voltage across the diode, you would find that the cathode voltage is negative compared to the anode.

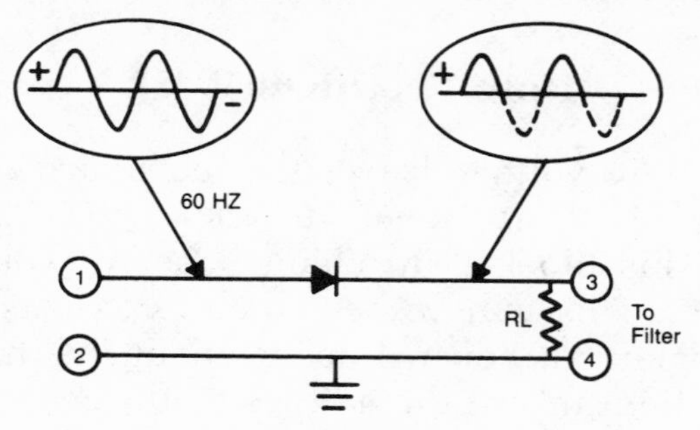

Figure 7–3
Diagram of Diode Action

Referring, again, to Figure 7–3, as the input AC waveform goes negative, the anode becomes negative with respect to the cathode. As you already know, this reverse biases the diode, and the diode quits conducting. With no conduction, there is no current flow and there is no power delivered to the load. This is where the term pulsating DC originates. During the negative half cycle, there is no current flow, so the voltage and current are delivered in a series of pulses. Pulsating DC is not adequate to deliver power to a circuit needing pure DC. An amplifier being fed pulsating DC would turn on and shut off 60 times a second, which would cause a 60-hertz tone that would be as loud as, or louder than, the signal being amplified. In order to remove the pulsation, the voltage is then fed to the filter, whose function is to purify the DC and filter out the ripple or pulsations. This will be explained further in the next section. If you desire a negative going voltage, you only have to reverse the polarity of the diode.

The diode will then conduct on the negative going half of the waveform, and cut off on positive going peaks. The simplicity of the half wave rectifier is one of its major advantages.

Simplicity usually means economy, which is the other major advantage of the half wave rectifier. There are major drawbacks, though. These include inefficiency, as compared to the full wave methods, which will be covered next. As you will see, the full wave rectifier uses the negative going waveform, and converts it to useful work in our supply, while the half wave circuit just ignores it. As you have probably already figured out, a half wave rectifier uses only half of the AC waveform, while the full wave circuit uses all of it. Another major disadvantage of the half wave rectifier is the need for larger filter circuits in high current supplies. Capacitors and chokes, which largely make up filter circuits, are expensive, and by using full wave techniques, the circuit may be less expensive than a half wave type.

From the foregoing, you may come to the conclusion that the half wave supply is not very useful; however, there are times when it is the best to use. You will learn which is the best for your application when you read the section on designing a power supply at the end of this chapter.

To achieve maximum efficiency, a rectifier should use all of an AC waveform, rather than just the positive or negative half. The full wave rectifier accomplishes this. There are two types of full wave circuits, the conventional, and the bridge rectifier. The conventional rectifier circuit contains two diodes. See Figure 7–4. Notice also that there is a different power transformer. This transformer has a center-tap, whose function is to reverse the polarity of the waveform to the lower diode. Referring to Figure 7–4, when the AC waveform is going positive in the transformer primary circuit, the waveform that appears between diode D1 and the center-tap is also positive going, while the waveform between diode D2 and the center-tap is negative going. Current flow in the circuit is in the direction of the arrows. Figure 7–5 diagrams the circuit and its current flow when the primary circuit contains a negative going voltage. As you can see, by comparing the two diagrams, when the waveform is positive, D1 conducts, while during the negative half cycle, D2 conducts. Because of the action of phase reversal, in the center-tapped transformer, the current

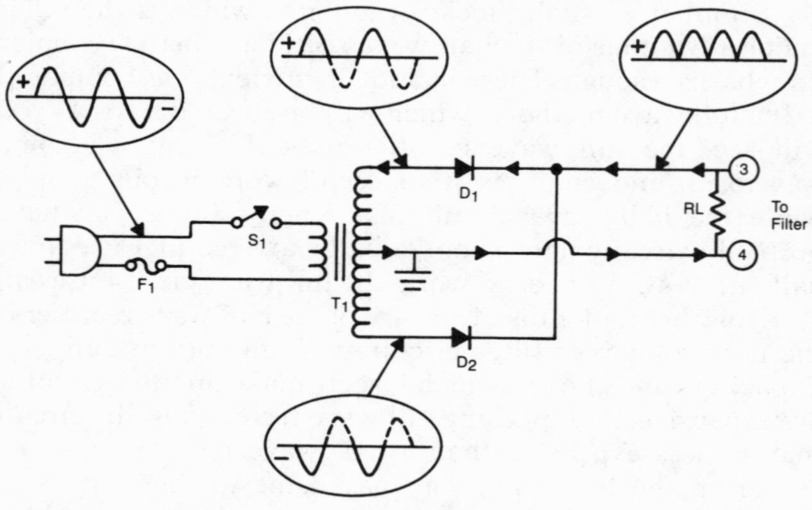

Figure 7–4
Conventional Full Wave Rectifier

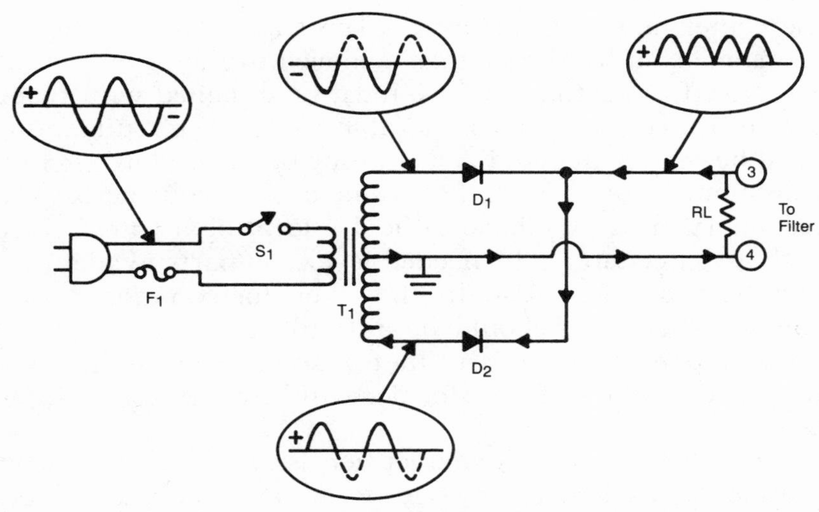

Figure 7–5
Full Wave Rectifier

flow through RL is always in the same direction. Notice also that the output pulses, (Figure 7–5) are closer together than the pulses in the half wave circuit. This results in a generally higher output voltage, and allows the filter circuit to be less complex, and thus less expensive. Observe that the DC output is still pulsating; there is just less time when the output of the supply rectifiers is at zero volts.

Typically, the output voltage of a filtered supply is higher than the AC voltage being put into the rectifier circuit. This would seem to be impossible; however, remember, as was stated, in the chapter on AC circuits, that AC is usually measured and the value stated in RMS. The AC peaks are 1.414 times greater than the RMS value. When the voltage is filtered, and delivering energy to the load, its actual DC voltage is near the peak AC voltage, rather than its RMS voltage.

The advantages of the conventional full wave are thus, the ability to utilize the entire AC waveform as useful output, and the lowering of requirements for filter circuits. There are some disadvantages to the conventional rectifier, which become apparent as you study Figure 7–5. The most obvious is the extra expense involved in the extra diode, and the center-tapped transformer. The slight disadvantage is overshadowed by the savings achieved in the filter circuit. The transformer that you use must be designed with twice the output voltage of a comparable half wave circuit, and it must be center-tapped. Each winding being used in our 13.8 volt supply must deliver 13.8 volts, or 27.6 volts across the secondary. You may have difficulty locating a transformer with a center-tap at the needed levels, so parts availability may be a problem. Usually, however, for commonly used voltage levels, you should be able to find an appropriate center-tapped transformer. In the section on designing a supply, the choice of transformer will be covered in greater detail.

The full wave bridge rectifier is probably the most commonly used rectifier style. It does not need a center-tapped transformer to function, so you may choose any transformer that has a winding that will output the voltage desired. At first glance, this circuit seems quite complex, however it is much easier to explain than it looks. Referring to Figure 7–6, current flows during the positive going half cycle through RL towards the junction of D3 and D4. Since current flows against the direction of the arrow, it appears current will flow through both diodes; however, since the anodes of the diodes are connected to opposite ends of the power transformer, the only diode that has a positive voltage on the anode is D3. D4 has a negative voltage on the anode, and is thus reverse biased, and will not conduct. Current takes the upper path through the forward biased D3. Power

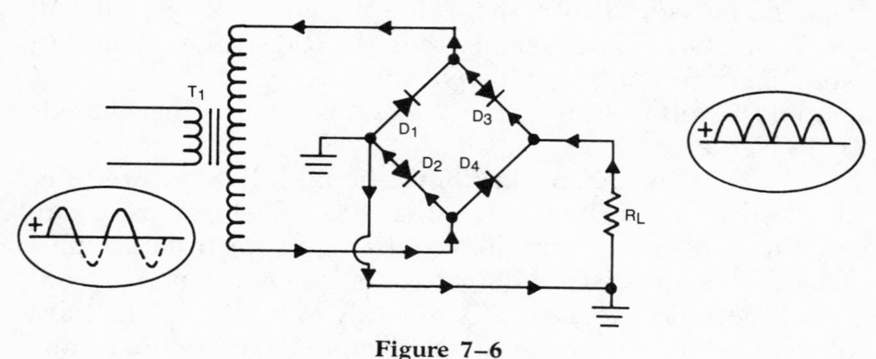

Figure 7–6
Full Wave Bridge Circuit

supply current then flows through the power transformer, and goes toward the junction of D2 and D4. Since D2's voltage is slightly positive on the anode, compared to the cathode, current flows through D2 and back to the load. Thus stated simply, during the positive half cycle, D2 and D3 conduct and D1 and D4 are off. As you have probably already concluded, during the negative half cycle, D1 and D4 conduct, and D2 and D3 are cut off (see Figure 7–7).

The full wave bridge circuit thus has all of the advantages of the conventional full wave, with the added advantage of not requiring a center-tapped transformer. A less apparent advantage is that since there are effectively two diodes in series, the reverse bias voltage across each diode is one half the secondary voltage; therefore, a less expensive diode, with a lower peak inverse voltage rating may be used. Also, for a given transformer, since the entire secondary is used, the voltage output of the bridge is twice that of the center-tapped conventional rectifier. To put this another way, you don't need a 28-volt transformer to get an output of 14 volts. The disadvantage, of course, is the need for four diodes rather than two, with no improvement in filtering requirements. This can be somewhat offset by the fact that a popular solution to this problem is to buy an integrated diode bridge. This is a circuit that has four diodes encapsulated into a single case with four leads, two AC inputs and a plus and minus DC output. These are available in several PIV and current ratings. More on this later.

The output of the rectifier block is connected to the

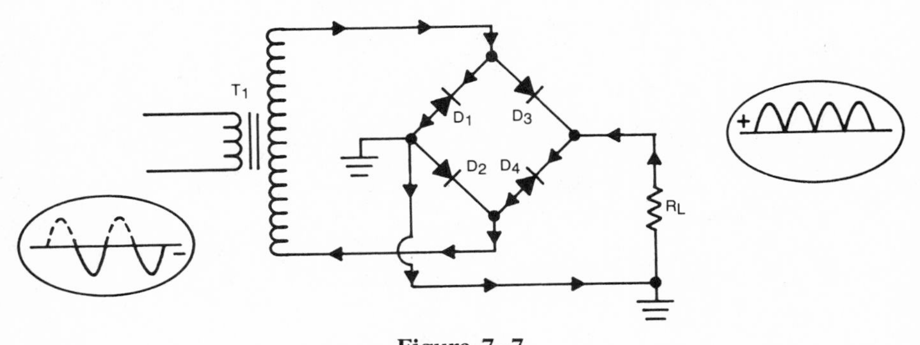

Figure 7–7
Full Wave Bridge Circuit

input of the filter block. This block is designed to remove the ripple voltage, or pulsating DC. This alternating voltage never goes negative (or positive, in negative going supplies), however since it is not pure DC it will not allow a circuit to work properly. This is the topic of the next section.

How Filters Work and Why They Are Needed

As we have seen, the rectified AC from the rectifier block is not even close to a pure DC signal. There are users for such signals; battery chargers and motor controls do not require a pure DC, for example. Most electronic equipment, however, requires properly filtered DC.

One of the most useful devices to the experimenter is the capacitor. As you remember, the capacitor has three basic functions. These are: to store DC voltages; to block DC voltages; and to allow AC voltages a path for current to flow. Refer back to Chapter 2 for a more detailed description of the capacitor. The power supply filter circuit uses all three of these functions to assist in the purification of the DC voltage. Let's see how this works. Figure 7–8A shows the output waveform from the rectifier in a half wave supply. The filter circuit shown (B) consists of only a capacitor, and resistor RL

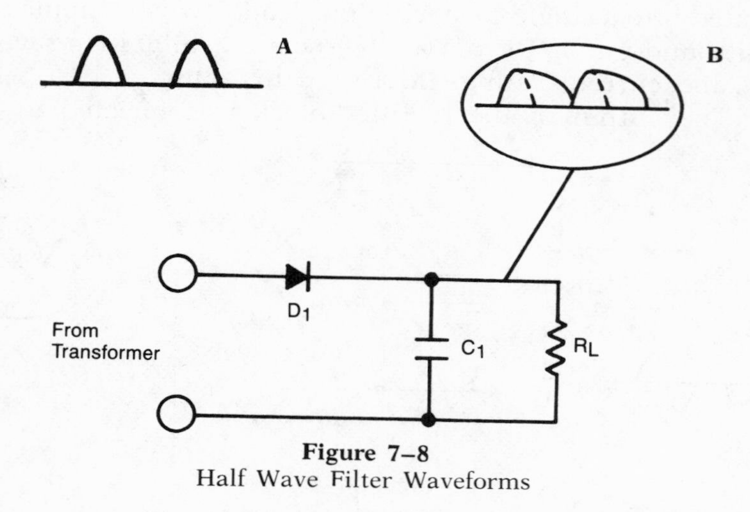

Figure 7–8
Half Wave Filter Waveforms

represents the load of our circuit being powered. As you can see by the output waveform in the circuit, the filter capacitor stores the DC voltage as the waveform is going positive. It blocks DC from travelling to ground, while allowing AC to do so. As the pulse heads toward 0 volts, the capacitor discharges its stored energy into the load resistor. As a result, the output is more like true DC. This circuit has a relatively low value of capacitance, so the voltage drops quite low. Figure 7–9 shows the output waveform of a full wave circuit. As you can see, there is much less ripple voltage.

Ripple voltage is the actual amount of residual AC that is present on the DC voltage. The lower the value of AC on the DC signal, the better the filter circuit. In the two circuits shown, the ripple voltage is rather high, and can be improved by either of two methods. One is to use a larger capacitor, capable of storing more energy during the positive going pulses, thus releasing more energy during the last part of the cycle. The other option is to increase the value of RL, which decreases the current required, therefore drawing less energy from the capacitor. The latter method is not always usable, since a circuit needs a certain amount of current to function. The point is that the lower the current requirements of your circuit, the easier it is to filter the AC ripple. As you have already seen, the full wave supply is much more efficient, since the frequency of pulses is doubled, and the capacitor is charged and discharged at a more rapid rate. This allows a given capacitor to be nearly twice as efficient.

The higher the current level demanded from the filter circuit, the poorer the filter performance will be. A good circuit design should have less than one percent ripple voltage. In other words, if the supply you are building is a 12-volt supply, it should have no more than 0.12 volts of residual AC at the load. Some circuits require better regulation than this, though there are circuits that are more

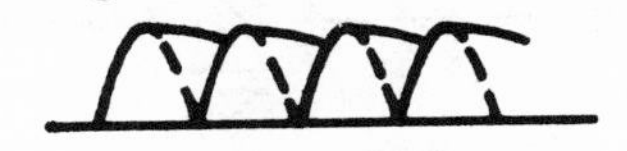

Figure 7–9
Full Wave Filter Waveforms

tolerant. The advantage of a single capacitor filter is its simplicity; however, to be effective the value of the capacitor must be quite large.

To increase the effectiveness of a filter circuit, the properties of the coil are utilized. Figure 7-10 shows a choke input filter. The action of the choke in slowing the change in current, combined with the effect of capacity showing the change in voltage, provides quite effective filtering action. This inverted L filter, named because of its resemblance to the letter L, is used when the circuit requires greater filtering action. Figure 7-11 shows the pi-network filter circuit. This circuit provides a slightly higher output voltage from the supply; however, it does so at the expense of a slightly higher ripple voltage than the choke input filter. These two circuits are commonly used in commercial power supplies, and the experimenter can design supplies using these filters relatively easily. There is a variation on the pi-network filter that is seen quite often. In order to save the high cost of a filter choke, some designers replace it with a resistor calculated to act as a resistive load. Although it is not as effective as a choke, if the demands of the load are not too stringent, the substitution is satisfactory.

The output of all filter circuits is dependent upon the load current required, as well as internal resistance of the rectifiers, coils, transformers, and even conductors in high current supplies. This internal resistance actually causes the output voltage to drop as current increases. For example, a supply that puts out 12 volts when there is no load, may only put out 11 volts when the supply is asked to deliver rated current. This characteristic of power supplies, called per-

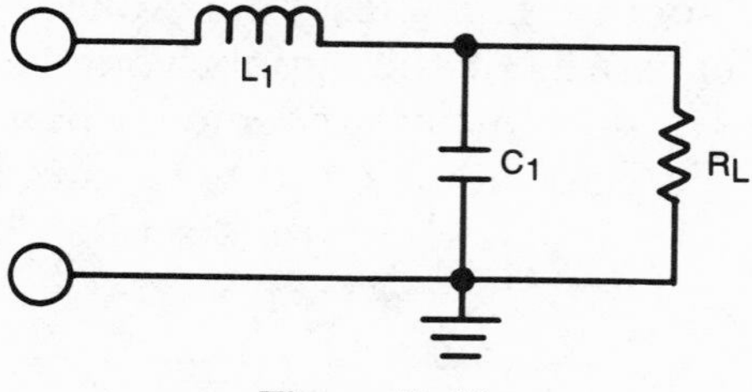

Figure 7-10
Choke Input Filter

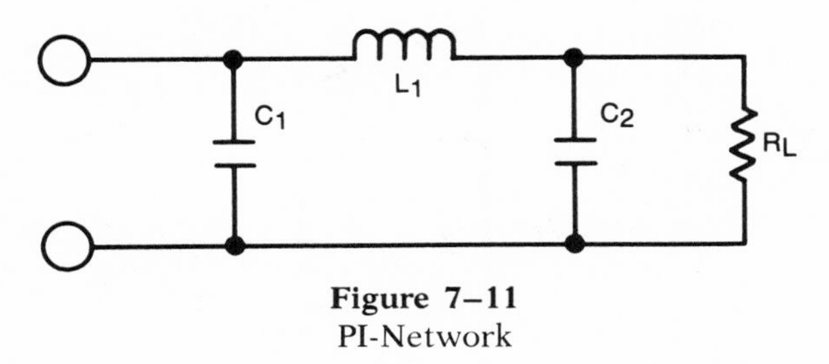

Figure 7–11
PI-Network

centage of regulation, is important to the experimenter. If the circuit you are building must have little voltage change, as the current delivered to the load changes, you will need a well-regulated supply. One way to decrease the effect of poor regulation is to design a supply capable of delivering much more current than will be required. This will insure that there will be little voltage change as current demands vary. This technique is often expensive, and makes the project larger and heavier than necessary, due to the increased size of the power supply. You might call this technique brute force regulation. With the advent of high gain transistors and integrated circuits, designers have devised techniques to improve regulation while using smaller size power supply components. Today three terminal regulators are plentiful, and inexpensive, and will improve the performance of even the simplest supply at a very low cost.

Regulators—What They Are and When They Are Needed

The voltage regulator is essentially an electronic filter circuit. Its purpose is to keep a constant output voltage regardless of the condition of the input voltage or changing load conditions. Electronic regulators usually consist of two or more transistors, and other components that allow the transistors to conduct in proportion to the changing circuit requirements. There are two basic types of regulator circuits, the shunt regulator, and the series regulator. Figure 7–12 shows both styles. Though the two regulators provide the same function, they do it in different ways (see Chapter 8 for more information on transistors.) In the series regulator,

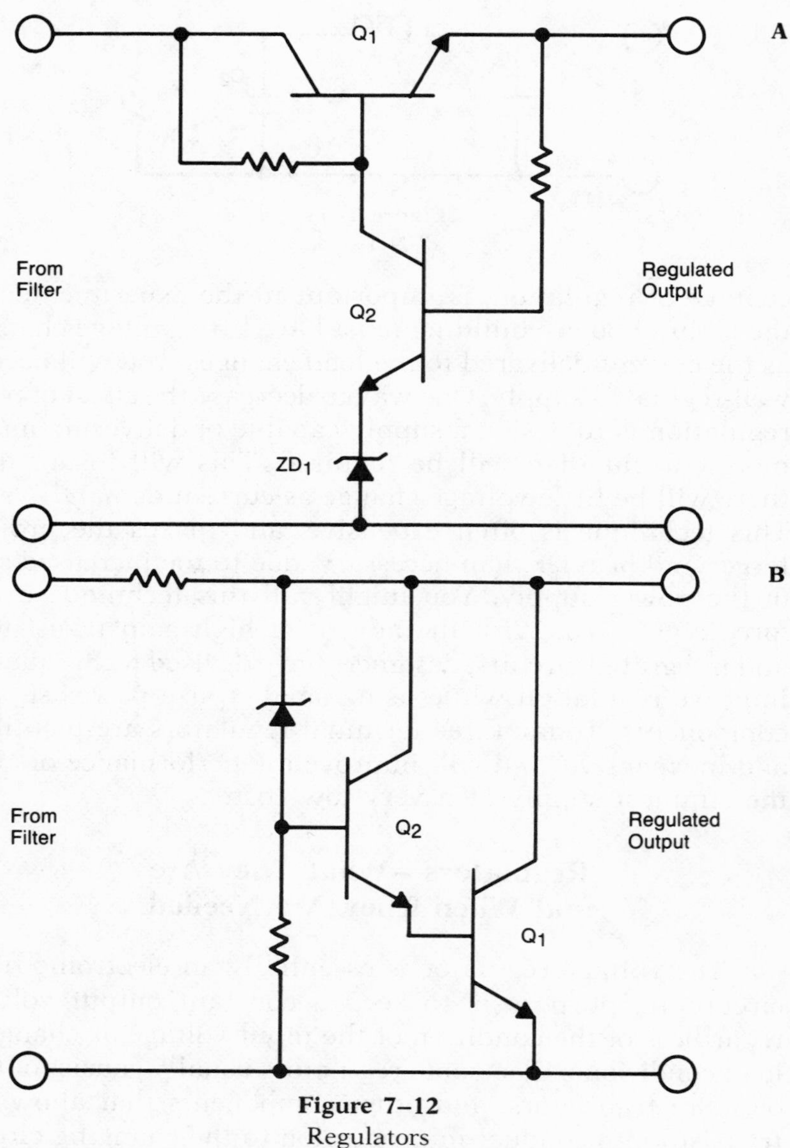

Figure 7–12
Regulators

Q1 is called the pass transistor. Its function is to provide the proper amount of resistance internally to keep a constant voltage at the regulator output. Let's see how this works. When the load requires more current, the voltage drop across Q1 increases, and the output of the supply starts to drop. This drop in voltage is sensed by Q2, the error amplifier transistor, via the bias resistors. The conduction of Q2 changes, causing more current to flow through Q1, thus lowering its effective resistance. The output voltage rises back to normal. The device continues conducting at this new level until a different current requirement at the supply again is sensed by the error amplifier. The zener diode, ZD1 provides a reference voltage for the circuit. Not only will the error amplifier sense a voltage drop due to changes in load current, it will also adjust its output when there is a change in line voltage at the input of the supply.

The shunt regulator operates in a similar manner; however, rather than lowering and raising the amount of current to the load through the pass transistor, it causes the current to be shunted or passed to ground through Q1, the shunt transistor. When a drop in voltage is sensed at the load by the error amp Q2, its conduction is changed, allowing the shunt transistor to conduct less heavily, shutting down the internal current in the device, and allowing more current to be delivered to the load. The zener diode is again a constant voltage reference for the circuit. With a shunt regulator, the power supply itself always delivers a constant current to the regulator and the load.

Though these circuits are easily designed and built, for most low current applications, there is an easier way for the experimenter to regulate the supply. This is the voltage regulator IC. These devices are available in several different current and voltage ratings. They can even be cascaded or connected with transistors to provide for higher voltages or currents if required. The advantage of these devices is in their ease of hookup, small size, and low cost. Information on selecting and using these devices will be given in the next section.

While it is most common to connect and use regulators in constant voltage applications, it is possible to design current regulators. Where a voltage regulator controls current to maintain a constant output voltage, a current regulator automatically adjusts its output voltage to maintain a constant output current. In other words, if a circuit requires a constantly controlled one-amp current flow, the varying load resistance is sensed by a current error amplifier. The regulator circuit varies the voltage delivered to the load in proportion to the amount of load resistance change. The net result is a varying output voltage and constant current being delivered from the supply. Constant current regulators are not typically used in most experimenters' projects, unless specific results are required for a certain project.

The experimenter must decide what voltage levels and current requirements are needed from a supply when building a project. Quite often, the power supply is one of the last circuits designed, as only then does the experimenter know the power requirements of the circuits. The next section shows how an experimenter might design a working supply to power his or her project.

How to Design a Power Supply to Fit Your Requirements

You have just finished a circuit that is designed to amplify the signal from your home stereo or cassette player to a higher power level to drive your speakers harder. You are ready to test it out and need a power source for the amplifier. If you designed the circuit yourself, you would probably have selected components that allowed the circuit to operate from the internal supply of the cassette player. More likely, either you found the booster amplifier in an electronics magazine and it was designed to be powered by the car battery, or, just as likely, the stereo you have does not have the current capacity to power any external circuits, especially high-power amplifiers. Now you are going to have to build a supply that will power the amplifier. At this point you must make several decisions about the use of the project, its requirements, and the power sources available. The steps to be followed in designing a supply are listed in order below, and explained separately following. All of these steps are interdependent, so the steps are listed in order, but the process is more integrated than the list implies. Consider each of the steps in order, and, as the concept of the supply changes shape in your mind, return to each step and consider it in the light of the new conditions.

1. Determine input supply voltage.
2. Determine supply voltage(s) and current(s) needed by the circuit.
3. Choose a suitable transformer.
4. Select rectifier diode(s).
5. Determine filter requirements.
6. Select regulator design if required.
7. Protect the power supply.

1. Determine input supply voltage. Are you going to use 110 VAC or 220 VAC? You might possibly be using some other AC or DC source. For example, your circuit might be for an automobile, yet need only 6 volts to operate. The power supply you design would then convert the 13.8 volt DC automobile supply to 6 volts. We will design a power supply that is to be powered by 120 volts AC.

2. Determine supply voltage(s) and current(s) required by the circuit. This is found by checking the circuit you have built for its operating characteristics. If you designed it, you will know how many volts and amps the supply must deliver. If you built it from a magazine article, the article will usually specify the power requirements, or at least the operating voltage needed. Once you have determined the operating voltage, you can always determine the current by using a bench supply to power the device. Insert an ammeter in series with the bench supply and read the operating current. For our experimental supply, we have determined that the required voltage will be 13.8 volts, and the maximum current the project will need is 1 amp.

When you are designing a supply for multiple voltage circuits, each step will detail the extra requirements. Total supply current delivered will be the sum of all of the currents at each voltage level.

3. Choose a suitable transformer. Not only will the transformer you select be dependent upon the current and voltage requirements, it will also depend upon whether or not the supply will be regulated. You must also decide whether to use a half wave, full wave center-tapped, or full wave bridge rectifier circuit.

Our example supply needs to deliver 1 amp of current, so a suitable transformer should be able to deliver a continuous 1.5 amps without overheating. This gives some margin for safety, and designers consistently rate their transformers at least this conservatively. Voltage requirements will vary with the need for regulation. Most regulator circuits need a margin of at least 2 volts in excess of the voltage to be delivered to the circuit. This voltage allows for proper regulator action. Since our supply is to be regulated, we can select a transformer that has a 120-volt primary and 15-volt to 18-volt secondary. These are commonly available, and can be obtained in various current ratings. Remember, when choosing a transformer, that the unloaded DC voltage will be slightly higher than the AC secondary voltage. It is possible to use a transformer of larger than the 1.5 amp rating specified earlier, should such a device be available at a flea market or in your parts box. The only disadvantage of using it might be its larger size. Also, if the supply is to be regulated, it is possible to use a higher-voltage transformer. The only caution is not to exceed the input voltage rating of the regulator device. More on this later. Unregulated supplies have to have transformers that are closer in tolerance to the output voltage required.

When designing a multiple-voltage supply, the transformer current rating must be 1.5 times larger than the sum of the individual source currents. The voltage rating of the transformer must be equivalent to the highest voltage desired. There are transformers available with dual secondary windings, and these simplify design requirements in calculating various output currents. The disadvantage of a multiple secondary transformer is that each winding will need its own rectifier assembly and filter circuit. They are best used when there is a large spread between voltage levels to be delivered by the supply.

In our design example, the required transformer will not need a center-tap, as we intend to use a full wave bridge circuit. Once the transformer is chosen, the next step is to select the appropriate rectifiers.

4. Select rectifier diode(s). Rectifier diodes must be chosen with the PIV and current ratings in mind. Diodes should have a 100% margin of safety. In other words, double everything. This may seem wasteful, however diodes are inexpensive, and an underrated diode can cause damage far in excess of its cost. When designing a bridge circuit, a lower current rating can be tolerated, as each diode is working at only half its current capacity. Another caution: Large filter capacitors demand high instantaneous charge currents upon power up. Choose a diode that has sufficient surge current capacity to withstand the high startup current of the supply.

In our example supply, we will use a bridge rectifier assembly. A readily available four-diode assembly has a 50 PIV rating at 2 amps. Any bridge assembly with specifications at least as large as these will work properly. If you notice any noise or static in your project that is not there when you try your project on your bench supply, the problem is probably noise generated by the action of the semiconductor diodes. To eliminate the noise, install a 0.001 mfd. disc capacitor in parallel with each of the diodes, or across the four bridge assembly terminals. See Figure 7–13.

When you are designing a multiple-voltage supply, be sure to add up the individual currents required by your loads. This combined total is then doubled to find the diode

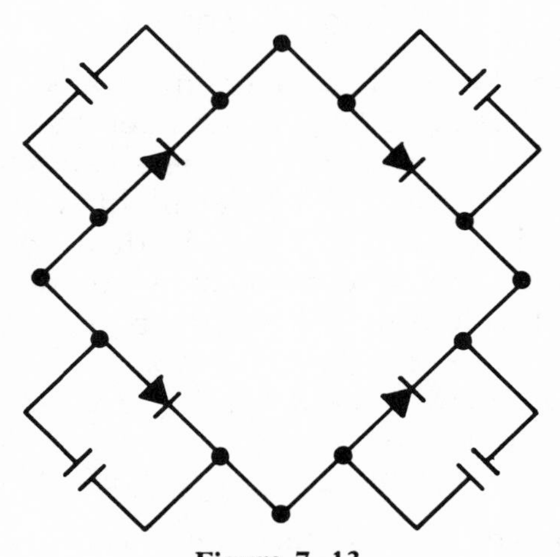

Figure 7–13
Bridge Assembly with Capacitors

current rating. Failure to do this can cause overheated rectifers and subsequent component damage.

5. Determine filter requirements. After deciding how much ripple is acceptable (1% is a good rule of thumb), you must choose a filter design. The most common choice is a pi-network type. To figure the amount of total capacity required, use your trusty calculator.

The procedure below will give an adequate filter design for most circuits. To calculate the input filter capacitor, two values must be known first. These are load resistance RL and ripple voltage Vr. The load resistance is the equivalent resistance of the circuit being powered, and is calculated with the formula

$$RL = Eo/IL$$

where RL is the load resistance, Eo is the output voltage of the power supply and IL is the total current the load will demand. The ripple voltage is easily computed with the formula

$$Vr = \%r\,(V)/100$$

where Vr is the ripple voltage, %r is the desired percentage of ripple, and V is the output voltage of the supply.

The final step in calculating the filter capacity is to insert these values into the following formula.

$$C1 = 0.00188/RL\,(Vr)$$

This will give the total capacity in farads. To convert to microfarads, just multiply the number by 1,000,000. The answer you get will probably not be a value that is readily available. Choose the next larger available value, or if that is within a couple of hundred farads then choose the next larger size yet. This is done because of the rather large percentage of tolerance in commercially manufactured capacitors.

In the example we are calculating, our one amp 13.8 volt supply, RL is found to be

$$RL = 13.8/1 = 13.8 \text{ ohms}$$

and the ripple voltage is determined just as easily by using

$$Vr = (1 \times 13.8)/100 = 0.138 \text{ volts}$$

To calculate C1, the input filter, the formula is converted to

$$C1 = 0.00188/(13.8) \times (0.138) = 0.0009872 \text{ farads}$$

Multiply by 1,000,000 and the value is 987.2 microfarads. The closest standard value is 1000 mfd.; however, since the desired value is quite close to this it would probably be better to choose a 1500 mfd. capacitor. The voltage rating of the capacitor can be determined by multiplying the output supply voltage by 1.5 and using the next higher value. A 25-volt DC capacitor will work just fine in our supply. In a simple power supply that is to be regulated, doubling the value of C1 would eliminate the need for a choke and filter C2. Since chokes are expensive, this is the course we will follow.

For pi-network filters the value of the choke can be approximated with the formula

$$L = RL/1000$$

while C2 can be calculated using the formula

$$C2 = 100/L(Vr).$$

One other comment on these calculations. These values and formulas are based upon the 120 Hz. ripple frequency of a full wave power supply. For a half wave supply, use the formulas as shown, and then just double the results. This simple trick will give a working circuit every time.

Multiple voltage supplies are more difficult to design. Refer to Figure 7–14 for the following. Assuming a typical example of a dual 12-volt and 5-volt supply, that is quite common in computer experimental circuits, calculate the design for the transformer and rectifiers based upon the higher voltage. To determine current, be sure to add the total current of all circuits. This total current is to be used in determining filter values. (See p. 192.)

At the output of the supply, add a dropping resistor (before the 12-volt regulator). If the lower voltage is to have its own regulator, a resistor is not always required. This resistor can be calculated by the formula

$$Rd = Vh\text{-}VL/IL$$

where Rd is the dropping resistor, Vh is the higher voltage and VL is the desired output voltage. IL is the circuit current

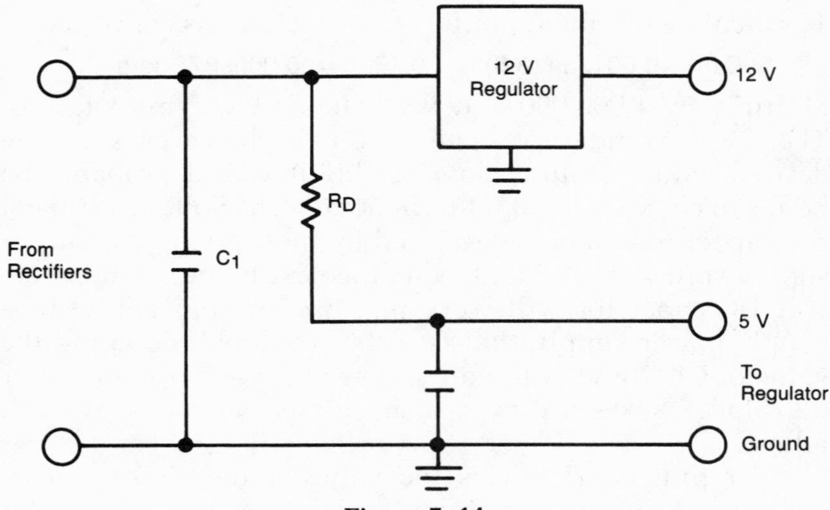

Figure 7–14
Multiple Voltage Supply

drawn by the lower voltage load. The wattage rating of the resistor is determined by

$$W = IL^2 \times Rd$$

where W is the power rating of the resistor in watts. Be sure that the lower voltage uses a relatively constant current. When using this method without a low-voltage regulator, the voltage drop across this resistor will vary directly in proportion to the current through it. A filter capacitor must follow each dropping resistor. Its value may be calculated using the formulas given to calculate C1.

6. Select regulator design if required. Although this is the last step, the choice of regulator determines earlier component choices. The transformer and filter values chosen are somewhat dependent upon whether or not you have regulated your circuit. With the low cost of three terminal regulator ICs and the relatively critical solid-state circuits often being powered, a regulator is often included. Low-current regulators are so inexpensive that much equipment designed today uses "on card" regulator ICs. On card simply means that, rather than building a high-current regulator to power several low-current circuits, each circuit is given its own regulator. This is the method to use, if possible, but it can't be done with a single high-current circuit. It is beyond the scope of this book to design high-current regulator circuits; however, a good source for this information is *The Radio Amateur's Handbook,* published annually by the American Radio Relay League.

The three-terminal regulator is so simple to install that you might think that there is no particular technique to follow. There are a few problems, however, that experimenters might run into. If you mount a regulator some distance from the final filter, install a 0.33 mfd. disc or mylar capacitor between the input of the regulator and ground, keeping its leads as short as possible. A large value output filter can be installed at the output of the regulator. It is not required, but improves the regulator's performance. Figure 7–15 shows these modifications.

Another common problem that the experimenter can run into is poor regulation of the supply. Often the experimenter believes the regulator is defective, and replaces it, only to find the problem still exists. This problem is usually caused when the supply voltage falls to within two volts or less of the desired regulated voltage. If this is happening, the chosen transformer does not have a high enough voltage rating. Also be sure to properly heat sink the IC. Regulators are chosen by their output voltages and currents, and also by the amount of input voltage they can handle. Our circuit

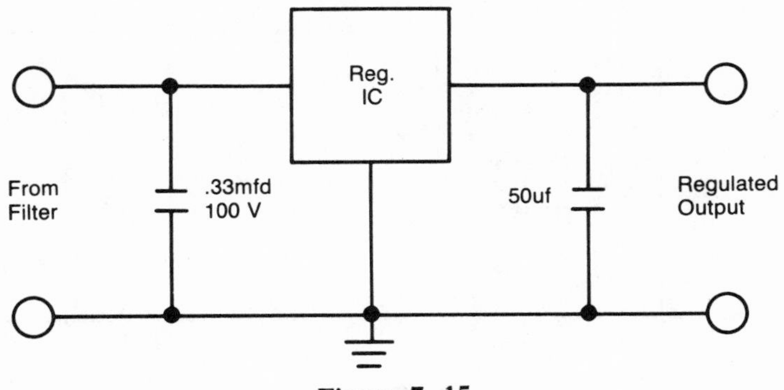

Figure 7–15
Regulator Circuit

requires a 13.8 volt at 1 amp regulator, however there are none available at exactly this level. There are two solutions to this problem, and they will be discussed separately.

Three-terminal regulators are usually available in 1.5 amp current ratings, so we will choose one of these. A popular regulator is the LM317K adjustable regulator. It can be easily adjusted by inserting a potentiometer in its control circuit, as in Figure 7–16. The other solution is to use a commonly available 15-volt regulator and insert two foreward biased diodes in its output lead. The foreward bias of 0.6 volts each will drop the correct 1.2 volts and we will have our desired 13.8 volts.

7. Protect the power supply. All power supplies should be fused in the primary of the transformer. To calculate the value of the fuse, use the following formula

$$If = ((Eo \times IL)/120) + 20\%$$

Other protection devices include crowbar circuits and foldback current limiting. The foldback current limiting is designed to protect regulators from shorts on the output of the power supply. Most three-terminal regulators incorporate this as part of their internal design. Crowbar protection is designed to protect the circuit being powered from regulator failure. Figure 7–16 shows a crowbar circuit. Zener diode ZD1 is a higher voltage than the circuit is

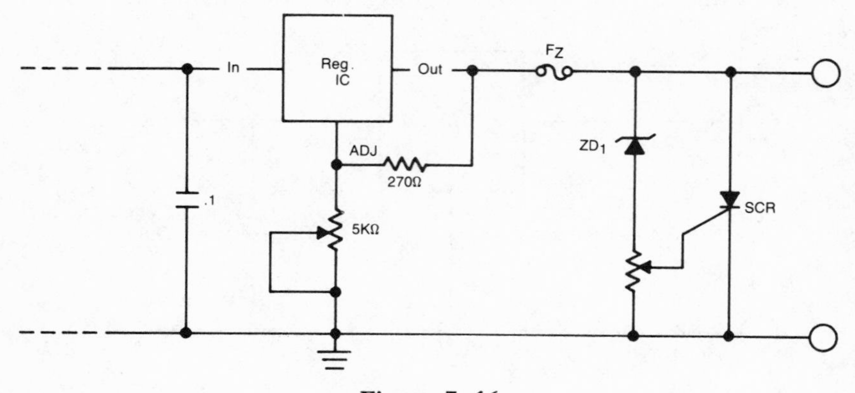

Figure 7–16
Adjustable Regulator with Crowbar

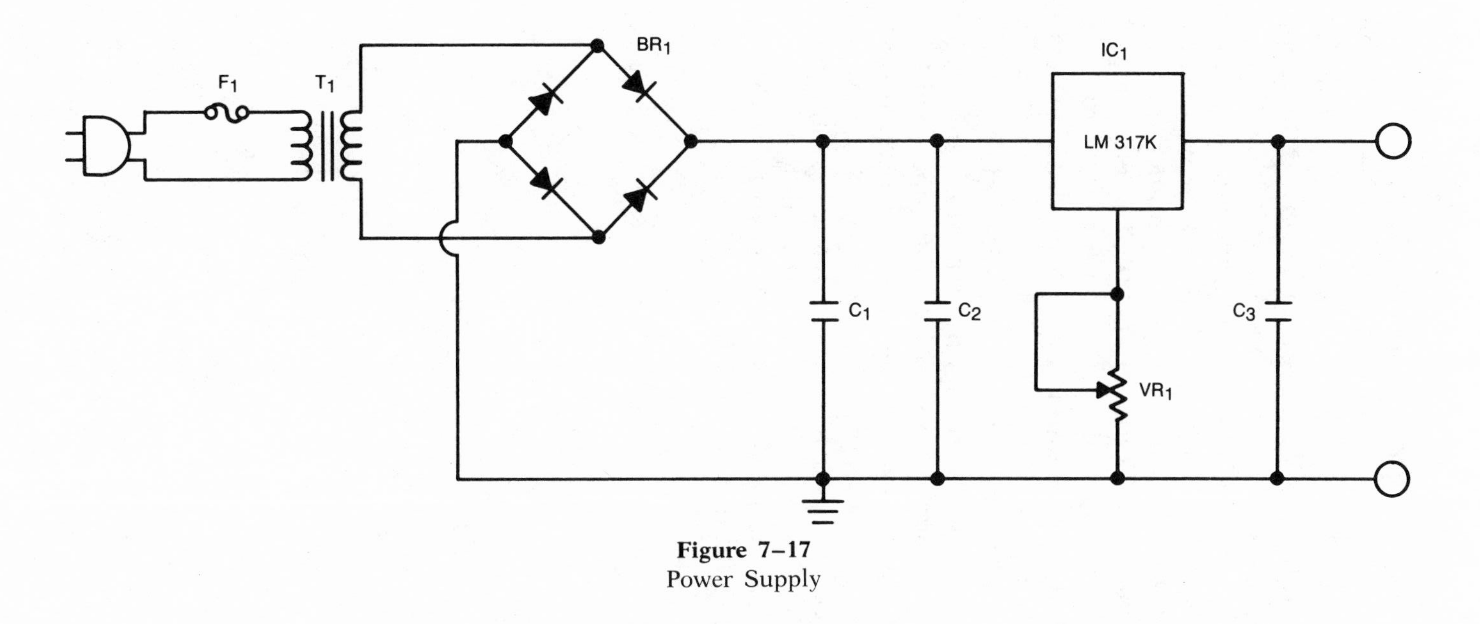

Figure 7–17
Power Supply

designed to put out. As a result, it is normally nonconducting. If the regulator fails and allows a high voltage into the supply line, the zener conducts, which in turn, allows the SCR to conduct. The SCR draws enough current to blow the fuse F2. The variable resistor adjusts the exact turn-on point of the SCR and is adjusted to just higher than the normal supply voltage. F2 must be installed, even though there is a fuse in the primary of the transformer. This fuse will protect the rectifiers from damage due to high surge current when the SCR conducts.

You will find, if you examine many power supply circuits, that there are many variations in design. Experiment with your supply to find out what works and what doesn't. One more hint on experimenting: Replace your circuit being powered with a resistor. Its value can be calculated by the formulas for RL and wattage given earlier. This will allow you to check and troubleshoot your supply without damaging your circuit. Figure 7–17 shows the final power supply design.

Transistors And How They Are Used

8

In the last chapter, we looked at power supplies and how they worked. One of the most important components we looked at is the diode. You saw that the rectifier diode has the ability to allow current to flow in one direction, while rejecting it in the opposite direction. Another component that has similar characteristics is the transistor. This chapter will demonstrate how easily transistor circuits are built, and how easy it is to make them operate.

You will also learn how to select, from the many hundreds of transistor types available, which would be best for your application. The best way for you to learn how to operate these devices is to acquire a few and check them out. This chapter assumes you can use the skills learned in Chapter 2, regarding the identification and testing of transistors. Take a few moments now, and review the sections on testing and identification of diodes and transistors.

The transistor, we will see, has three major applications—switching, amplification, and oscillation. Each of these processes is discussed separately.

How the Transistor Works

As you recall, the transistor is a sandwich of two semiconductor junctions. The transistor is basically an "off" device. The experimenter must know how to cause the transistor to conduct, or "turn on." When working with the diodes in Chapter 7, you will recall that a diode must be forward biased before it will conduct.

In transistor circuits, there are two conditions that must be met before conduction will occur. Basically, as the circuit in Figure 8–1 demonstrates, the emitter-base junction must be forward biased, while the collector-base junction must be reverse biased. Notice that the battery between the emitter and base is connected with the negative lead toward the emitter, and the collector battery is connected with the negative lead to the base. Looking at the block diagram of the transistor at the top, you can see that it is an NPN transistor. The emitter battery is connected with the N material to the negative battery terminal, while the positive battery terminal is connected to the P material. This meets the definition of forward bias. Let's see what happens to the transistor, when we connect it as in the figure. Do not try to wire these circuits as yet; they are simply for identifying the operational characteristics of the devices. There will be experimenter circuits that you can wire.

Figure 8–2 demonstrates what happens inside the device. All circuits will be demonstrated using an NPN transistor. Using PNP devices requires only the reversal of

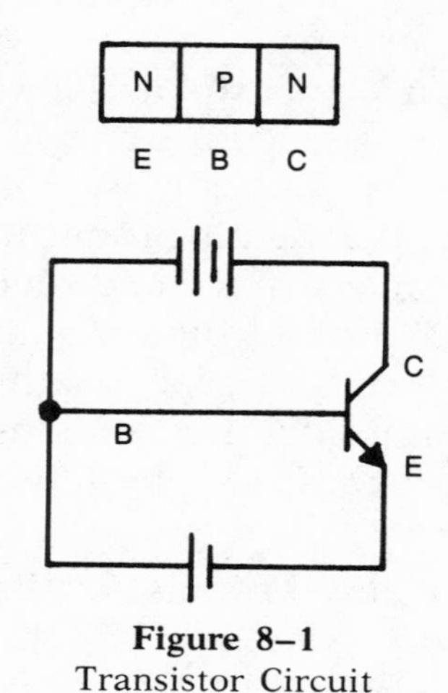

Figure 8–1
Transistor Circuit

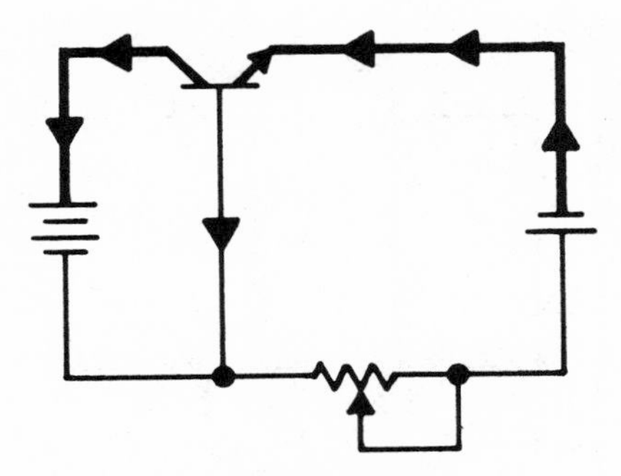

Figure 8–2
Transistor Circuit Internal View

battery polarities. More details of this will follow. Current starts to flow through the emitter and base junction. The high positive voltage on the collector terminal pulls electrons through the collector junction, and into the collector battery. Approximately 97 percent of the electrons travel into the collector circuit, leaving only 3 percent to travel into the base circuit. By adjusting the potentiometer in the base circuit, you can vary the amount of base current, which also changes the amount of collector current flowing in the device. Even though this circuit would work, it is not very practical. For one thing, it requires two power sources. Let's redraw the circuit so that it requires only one battery. This is a more practical circuit, and one that requires little special care.

Figure 8–3 is a circuit you can wire using a few locally obtained parts. The transistor is a 2N2222. If you do not know how to locate a transistor, or a suitable substitute, return to this point after reading the next section. You might also ask for assistance at your parts supplier. The light emitting diode, LED1, is a general purpose diode with a current rating of at least 50 ma.

Wire this circuit on a protoboard, or perf board, then check to be sure that the transistor is connected correctly into the circuit. When it is complete, adjust the potentiometer, Ra, to minimum resistance. If the circuit is wired correctly, you will be rewarded by a glowing LED. The

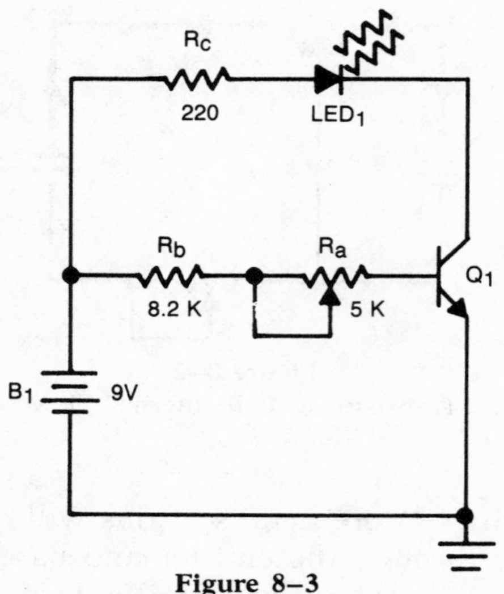

Figure 8–3
Transistor Circuit

circuit was designed to deliver enough current to light the diode, and in the next section, you will learn how to calculate for those values of bias resistors, so you can design your own circuits.

Slowly adjust the potentiometer, and you will find that the brilliance of the LED changes. As you increase the resistance in the base circuit, you will reduce the current flow in both the base and collector circuits.

To see the effects of what is happening inside the circuit, insert an ammeter into the collector circuit, either between the diode and transistor, or between the diode and Rc. The meter should be set on the 50 ma. scale. Energize the circuit, then while varying the base resistor adjustment, note the collector current readings. After noting the values of maximum and minimum Ic (collector current), remove the ammeter, and install it in the base circuit, between Ra and Rb, or between Ra and the base lead. The meter must be on the 1 ma. scale. As you adjust Ra, you will find that current in the base circuit will vary, but will be at least fifty times smaller than the collector current. You have just seen the major advantage of the transistor. Very small currents in the base circuit control relatively much larger currents flowing in the collector circuit. Practical applications for this will be shown in the last three sections of this chapter.

As was stated earlier, if you are working with PNP devices, you have only to reverse the battery polarity to allow the transistor to operate. As you can probably guess, the transistor junctions must still be biased exactly as they are for an NPN. To meet these requirements, just reverse the battery polarity.

Engineers and technicians use the terminology of "hole flow" to explain the operation of PNP transistors. If you run into this term from other sources, you will know what they are talking about. All transistors have majority and minority current carriers. In an NPN transistor, majority carriers are electrons, and we can discuss the operation of the device using terminology we are familiar with. The PNP devices, however, have "holes" as majority carriers. By now, you are probably asking just what are these holes. The holes are spaces in the semiconductor where electrons would fit. An easy way to think of hole flow is to imagine that if electrons are moving in one direction, the holes they leave appear to move in the opposite direction.

One question often asked is why are there two types of transistors, when only one would do. Today, indeed, most circuits use NPN devices, because they are easily connected to the negative ground circuits so commonly in use. PNP devices, however, are extremely useful any time a circuit requires a positive ground. Another example of the need for both types is in amplifier circuits. Some amplifiers use complementary symmetrical output circuits. A more detailed explanation is found in the section on amplifiers. For now, the amplified waveform is sent to both transistor types, where the part of the waveform that has the correct polarity is amplified by the appropriate transistor.

Before we look at using a transistor in a circuit, let's take a look at exactly how we can choose a particular transistor for a given application. Several important transistor parameters are discussed, and when you have completed the next section, you will be able to select a given transistor for a certain property, or group of properties.

How to Select a Transistor for Your Application

This section discusses exactly how you can select a transistor for a specific purpose. As was stated earlier, the three basic purposes are switching, amplification, and oscillation. When choosing the device, you must know the purpose. Another objective of this section is to show you how to select a suitable substitute transistor, given only a type number and a cross-reference guide. Also included is a complete description of the five most important parameters, or characteristics, of the bipolar transistor.

As an aid in learning about transistors, a semiconductor cross-reference guide is invaluable. These manuals, distributed by semiconductor manufacturers, contain a wealth of information on that company's line of semiconductor products. Also included in most guides are diodes and integrated circuits.

Due to the tremendous amount of semiconductor devices on the market today, there is very little standardization of part and type numbers. About the closest that the industry has come to standardization is the implementation of JEDEC numbers. Semiconductor devices using JEDEC numbers are easily identified. Let's look at a couple of examples. 1N914 and 2N2222 are both JEDEC numbers. The 1N signifies that the component is a diode, while a 2N signifies a transistor.

Unfortunately, very few companies use these numbers on their parts. They either number them with their own numbers or in some cases, they don't number them at all. In the latter case, there is very little that can be done to replace the device, short of running some very expensive test procedures. For the average experimenter, unmarked transistors can be considered not replaceable. If you must replace an unmarked device, there are a few things you can check, however, and with research, you might just be able to locate a reasonable replacement. This is especially true in switching and low-frequency applications. For now, let's look in further detail at the cross-referencing guide.

An excellent example of a substitution guide is provided by the Phillips ECG corporation. It lists several thousand semiconductor devices, and crosses them to over a couple of hundred ECG replacement devices. As a bonus, these devices are listed with all of their important electrical specifications and physical dimensions. In addition, the front of the manual contains several pages on semiconductor testing and replacement. Other manuals from other manufacturers, such as GE, RCA, Radio Shack and many others, are also available. To locate one of these guides, check with your supplier of electronic components. They may even be able to obtain a complimentary copy for you. You should obtain a guide for that brand of semiconductor devices most readily available to you.

To use one of these guides, let's try an example. The transistor used in the last circuit was listed as a 2N2222. You must look in the large section that contains those long, long lists of transistor part numbers. Each column lists two part numbers. The left number is the desired device, and the right column lists the replacement device that can be used. The reference guide lists the number 212 next to the listing by the 2N2222. This means that you should order a number 212 device, and it has equivalent operating characteristics to the 2N2222. By the way, if there is a single letter following the number, e.g. 2N2222A, and you find the first part of the number, but not that letter, you can usually use the transistor that replaces the one in the guide. That last letter is usually a designation of an improved device. Its basic characteristics are usually quite close to those of the original, or it would probably have a totally different number.

It is all well and good that we can replace a device that is in the cross-reference book, but what about a device that doesn't happen to be listed? (How can that be, with all those numbers?) To replace a transistor like that, we have to know about some of the characteristics of that particular device. Let's look at each of the important specifications and see what kind of information they provide.

Ic (collector current): The amount of current that can flow through the collector element before the device overheats and destroys itself. This rating can range from a few milliamperes for small signal transistors to several amps for large power transistors. Select a transistor with a collector current of at least 50% larger current capacity than you know the circuit will deliver.

Ib (base current): The maximum amount of current that can flow in the base region. Again, if this number is exceeded, the device will be destroyed.

Vbco (voltage between base and collector with emitter open): This rating is the absolute maximum voltage that can be applied between the collector and base. If this value is exceeded, the transistor will be permanently damaged. Note: The "o" on the last part of the specification signifies "open," and means that the missing lead, in this case the emitter, is open and not connected to any circuit.

Vebo (voltage between emitter and base): The absolute maximum difference of potential between the emitter and base. If this value is exceeded, the base-emitter junction will be destroyed.

Pd (power dissipation): The ability of the collector element to radiate or conduct heat away from the device. This rating can be increased by installing a heat sink on the device. As a result, many reference guides list two specs for Pd. One is for free air dissipation, the other is listed with an adequate heat sink.

There are over 150 different specifications relating to transistors that circuit designers may have to take into account. However, these five specifications will assist you in replacing just about any device you might need to obtain. Since all of the above specifications are absolute maximum ratings, identification of those ratings on the device you wish to replace is all that is required. Just choose a device that has ratings that are equal to or higher than those of the device being replaced.

There is one more specification that you will need for many circuit designs, and that is the upper frequency limit. The upper limit is defined as the point at which the ability to amplify drops below a certain point. Compare the requirements of the desired circuit with the replacement transistor in question. For example, an amplifier circuit you are building needs an upper response of 1 MHz. The transistor you choose must have a high-frequency limit of at least 1 MHz. This is one rating that must stay relatively close to the absolute value. In other words, don't buy a 5 MHz. transistor for this application. As you will soon see in the section on oscillators, the transistor would be very efficient, and that can cause some unexpected problems. For now, just remember that the device should have an upper limit of no more than two times the upper frequency expected to be used.

Now that you know how to use a substitution guide, look up a few transistor and integrated circuit numbers from the projects in Chapter 11. This will give you some practice at using the guide. For the transistors, note those important absolute specifications mentioned above. In the next section, we will look at the requirements of transistor bias circuits, and some more practical transistor circuits.

How to Bias a Transistor

As was said earlier, the transistor must have a forward biased base-emitter junction, and a reverse biased base-collector junction. Let's see just exactly what this means. Figure 8–4 contains a simple transistor switch. For now, don't worry about how the device switches, we will look at

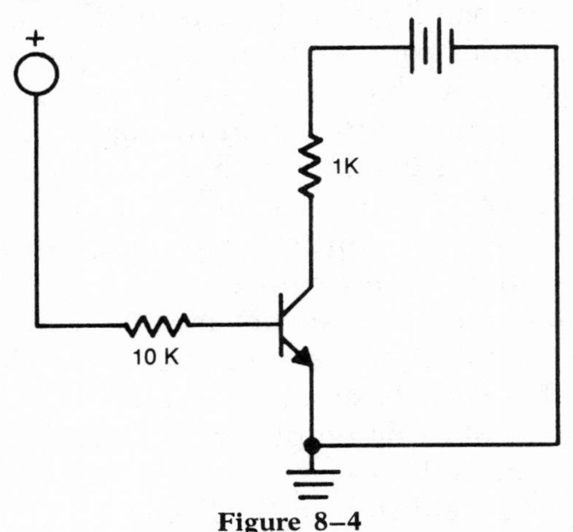

Figure 8–4
Transistor Switch

that in the next section. Notice, though, that the transistor is an NPN device. A forward biased junction should have the negative voltage on the N material, and a positive voltage on the P material. Since this is an NPN device, the base should be more positive than the emitter. This is indeed the case. The collector must be more positive than the base. At first this might seem backwards, as the collector is N material, however you will realize it is correct if you remember that the base-collector junction must be reverse biased. Simply, this means that you can immediately tell the polarity of any given bias circuit by just looking at the middle letter of the transistor. For an NPN transistor, the base must be more positive than the emitter, and the collector must be more positive than the base. Using your voltmeter, check the circuit you wired for Figure 8–3 to see if it meets those requirements. For a PNP device, you have probably already guessed that the N base lead requires a negative voltage supply, just as we have already stated.

One of the most interesting characteristics of the transistor is its versatility. There are three commonly used ways of connecting transistors to operate. Each way of connecting the device is used to meet certain requirements of the circuit designer. Figure 8–5 diagrams the three circuits, called "common base," "common collector," and "common emitter." Since a transistor has three terminals, one terminal must be common to both the input and output of the device. All of the circuits we have seen so far have been common emitter devices. The emitter is, in this case, grounded, and more important, it is common to both the input and output of the device. The input of a common emitter device is between the base and emitter, while the output is taken between the collector and emitter.

A common base transistor, as you can see by the diagram, has its input between the emitter and base, while its output is between base and collector.

A common collector amplifier (also called an emitter follower) has input between the base and collector, and output is between emitter and collector.

Each basic circuit type is used for different reasons, and designers select the appropriate mode for the particular

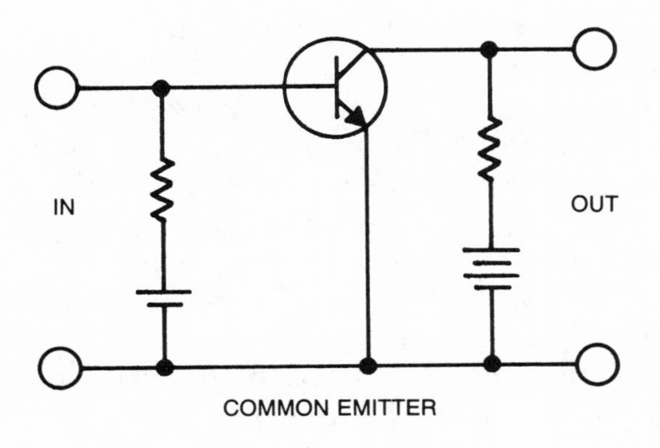

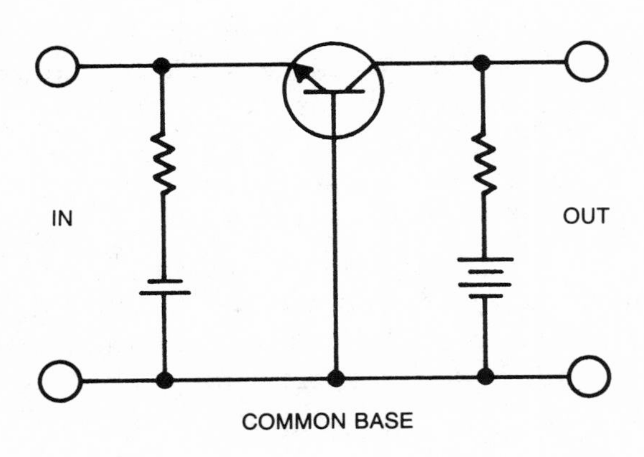

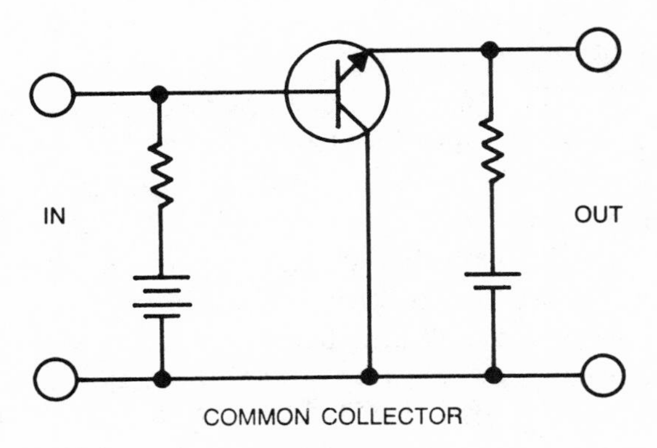

Figure 8–5
Common Base, Collector, Emitter Circuits

application. For example, the common collector amplifier is very weak in its ability to amplify, but it is commonly used to match impedances between two poorly matched circuits. As you probably recall, circuits that have matched impedances operate most efficiently. As a result, the extra power drawn from the emitter follower is more than returned by a much more efficient design.

The next three sections describe the principles involved in making transistors perform any of their three functions. All modes of connection require power and bias is the name usually given to the DC operating voltage supplying power to the device. There are several classes of bias, known to experimenters as class A, class B, and class C. There is also a sub-group, class AB. Each class has different requirements of the power supply, and each has a different efficiency. All of the projects in this book use class A amplifiers, though a couple of IC projects use class B circuits. Let's see just exactly what is the definition of each class.

Class A: A transistor operating in class A bias conducts continuously, from power up, until all power is removed. It is the least efficient form of amplifier, but the fidelity of amplification is excellent.

Class B: The class B amplifier conducts only one-half the time of a class A. In other words, it will amplify only the positive (or negative) half of an AC waveform. It doesn't sound too useful, but the section on amplifiers has a couple of practical examples of exactly how useful a class B amplifier can be.

Class C: The class C device operates, and takes current from the power supply, less than fifty percent of the time. The output of a class C amplifier compared to its power input is favorable; in other words, the circuit is very efficient. You can probably guess that it does not have very good fidelity of reproduction. The section on amplifiers points out when class C amplifiers are used, so I will defer further comment on their application until then.

Any device can be biased to operate in any of the above classes; it is up to the experimenter to determine when and why he or she wants to select a certain bias class. Switches, amplifiers, and oscillators all have different bias requirements, depending upon their application—and the next sections discuss the proper selection of bias and calculations of bias values.

The next section specifically looks at the transistor switch. The biggest reason for using a transistor switch is that the control circuits contain very little current, thus allowing long control runs, while the transistor switches the very large currents. More details are coming, along with some practical design examples.

How to Make a Transistor Operate as a Switch

We have talked a lot about how a transistor works and we have used one in a circuit, but we haven't really put one to work. That is the object of the next three sections. Before we get into the actual circuit characteristics and construction, let's look at some applications. The transistor switch is a commonly used device, yet many people don't understand exactly why a standard switch cannot be used. A transistor switch is used in many cases, when a relay or direct switching isn't practical. Relays are used to provide high-current switch contacts, close to the circuit being switched. They are controlled by low-current coil lines that may have to travel a long distance to the control point. Since the control wiring is small in gauge, the wire is less expensive, and the actual switch itself can be lower in current rating.

There are a few terms that must be defined before we can explain the operation of the electronic switch. These terms refer to the actual operating condition of the transistor or electronic switch. "Cut-off" is a term used to describe a transistor that is not conducting. In other words, it is in the off state. "Active" is the term used to define a transistor that is conducting, or drawing current between the collector and emitter. "Saturation" is used when the transistor is conducting as hard as it can. Transistors that are saturated are said to be fully on. As we progress through these devices, each of these terms will be explained as the circuit is designed and wired.

Before we design a transistor switch, let's take a look at one that has already been designed. We can tell a lot about exactly what is happening in the circuit by analyzing an operating device. Take a look at the circuit of Figure 8–6. It is almost identical to the LED driver circuit that was built earlier, but it is missing the variable bias potentiometer. The potentiometer has been replaced by a simple SPST switch. Wire this circuit on a breadboard or perf board and let's analyze its operation. SW1 can be any simple single-pole single-throw switch, push button or toggle.

Begin by measuring a few voltages, and see exactly what is happening. First, place your voltmeter negative lead at the ground lead of the transistor (which goes to the minus lead of the power source, a 9-volt battery). Next, to be sure the meter battery circuits are working, and connect the positive meter lead to point A. If all is well with the battery, the meter should indicate 9 volts.

The first step in understanding solid-state switch operation is to see the relationship between the input circuit and

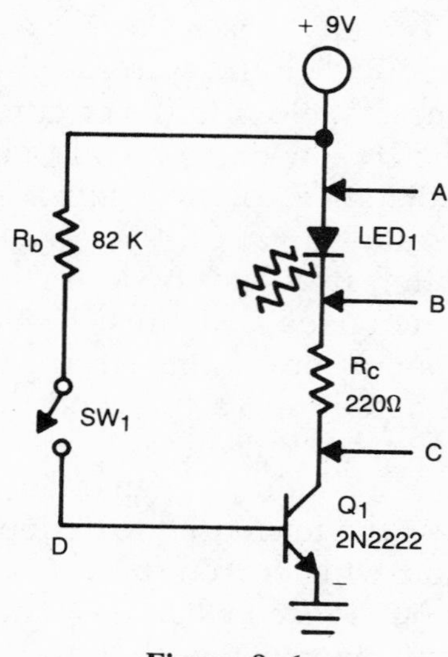

Figure 8–6
Transistor Switch

the output of the device. Since the switch is off, no current is being delivered to the base lead. Since no current can flow in the base circuit, any current flowing in the collector circuit is leakage current. A 2N2222 should have little or no leakage. Because the device is cut off, there is no current flow between the emitter and collector. That being the case, the light-emitting diode is off. If you measure the voltage between ground and the diode, you will find it to be very close in value to the supply voltage. Ideally, they should be identical.

When you switch the device on, the battery voltage is applied to the base bias resistor and current flows in the base circuit. If all is well at this point, the first thing that you should notice is that the LED should be lit. Now measure the voltage between the base lead on the transistor and ground. It should measure approximately 0.6 volts. Now measure the collector voltage (the voltage between the collector terminal ground) on the transistor. It should be at or near zero volts. From Kirchhoff's law, remember that resistors with low resistance have small voltage drops across them. Since the device is now saturated, its internal resistance should be at or near zero ohms, therefore the voltage drop across it should be also at or near zero.

You may find that the transistor you are using might not actually saturate. You can tell if this happens when you measure the collector-to-emitter voltage. If this voltage is above 0.6 volts and less than the supply voltage, the device is indeed conducting, however it is not saturated. Saturated transistors will have a lower voltage on the collector than on the base. For our purposes, if this should happen to you, for now, just decrease the value of the base bias resistor a few hundred ohms at a time until the collector voltage falls to zero. Later, you will learn how to calculate these values for yourself.

The object, as you can see by this project, is to see how the transistor switch operates. When no base current is allowed to flow, there is no collector current, and the switch is off. When base current is applied, the transistor is allowed to saturate, and its internal resistance falls from infinite to zero ohms. You have seen how easily these electronic switches are wired, and now we will look at exactly how the values of bias resistors are calculated. To do this, there is one more specification with regard to transistors that you must know about.

One measure of efficiency of just about any device, is a measurement comparing the output of a device to its input. All transistor circuits—switch, amplifier, or oscillator—have several different ways of rating the individual device. In transistor common emitter circuits like our switch, this rating is known as beta; the symbol is the greek letter β. It is easy to calculate the beta of a given device. You can do it with the circuit you have already built. First, measure the base current of the transistor, then measure its collector current. Beta is calculated by the formula Ic/Ib.

Beta must be determined before you can calculate any other component value in the circuit. One specification that is included in most transistor specifications is its beta. Unfortunately, two different transistors of the same part number often have two different values of beta. The number found in specifications is usually a nominal or average value, and component values must be "fine tuned" to the individual value of beta. In amplifiers, this requires a more elaborate bias circuit; however, for designing switches, an adjustment of the calculating formula is all that is required.

Now we are ready to begin designing our switch. Let's look at each of the seven simple steps one at a time. Two different examples will be used, which will demonstrate the versatility in designing the switch to fit your application. As a first example, let's see how that original switch you built had its values calculated.

Step 1. Define the load. In this case, the load is the LED, which we wish to turn on and off. The load current requirements are specified with the particular LED you happened to choose. This is why your device may not have saturated, as the requirements of the LED you chose may have been different from ours. The LED we are using will give adequate output at 50 ma. with a voltage drop of 1.8 volts across the device. The collector current of our transistor, when the switch is on, must not exceed 50 ma.

Step 2. Specify output power supply voltage. Essentially, this is the voltage that will appear across the transistor, when the switch is off. Typically, this will be a battery or power supply voltage that is readily available.

Step 3. Calculate collector resistor (if needed). This resistor is not required unless the supply voltage is greater than the voltage of the item being switched. In this case, the LED requires only 1.8 volts across it. We want to use a nine-volt battery to power the circuit, however, so a current limiting resistor must be added to the circuit to protect the LED. This value is easily calculated by applying Ohm's law. First determine the voltage drop required across the collector resistor.

$$Er = Et - El$$

The voltage drop across the collector resistor is determined by subtracting the voltage across the LED from the source voltage. (This assumes that the voltage drop across the transistor is going to be zero, as it should be if the device is saturated.) In our example,

$$Er = 9 - 1.8 = 7.2 \text{ volts.}$$

Now to calculate the value of the collector resistor. Applying Ohm's law, we find the formula to be

$$Rc = Er/Ir.$$

The collector resistor value is calculated to be

$$Rc = 7.2/0.05 = 144 \text{ ohms.}$$

Use the next closest standard value, in this case 150 ohms.

Step 4. Determine input voltage. This is the voltage that will be applied to the base resistor. It may or not be the same as the collector supply voltage. In our example, it is the same as the collector supply, as it also comes from the nine-volt battery. An example of another source could be a five-volt input from a digital IC.

Step 5. Specify transistor type. Now we have to determine the part number of the desired transistor. The specification guide is used to choose a transistor based upon the following criteria.

A. Maximum Ic of the transistor greater than Ic to be drawn when switch is on.

B. Maximum Vceo to be greater than the supply voltage.

C. Frequency response greater than the expected switching speed.

(Remember, the switch may be operated by digital logic circuitry, several thousand times a second if desired.) These specifications are all that are necessary to determine the fitness of a particular transistor in a switching application. Use any other criteria to fit your application, eg. case size, cost, availability, etc. In our application, we specify a 2N2222.

Step 6. Calculate base current. Earlier, it was pointed out that beta is equal to collector current divided by base current. By rearranging the formula, we can use the transistor beta in the spec sheet and the desired Ic to determine the actual Ib.

$$Ib = Ic/beta$$

Since the value in the spec sheet is nominal, to ensure that the transistor you purchase actually saturates, double the value of Ic when making the calculation. In our example,

$$Ib = .1/100 = 0.001 \text{ amps.}$$

The nominal beta of a 2N2222 is 100, and desired Ic is 50 ma.

Step 7. Determine the value of Rb. The final step is to apply Ohm's law to determine the value of Rb.

$$Rb = (Vin - Vbe)/Ib$$

where Vin is the base input voltage specified earlier, and Vbe is the voltage drop across the transistor base and emitter junctions. Remember Vbe for a silicon device is 0.6 volts, while a germanium Vbe is 0.2 volts. In our example,

$$Rb = (9 - 0.6)/0.001 = 8400 \text{ ohms.}$$

Use the next closest standard value.

Let's look at another example. This time we are going to use a transistor to switch a relay. This type of circuit is called a relay driver, and is often used when digital circuitry must switch high-voltage circuits. Figure 8–7 is an example

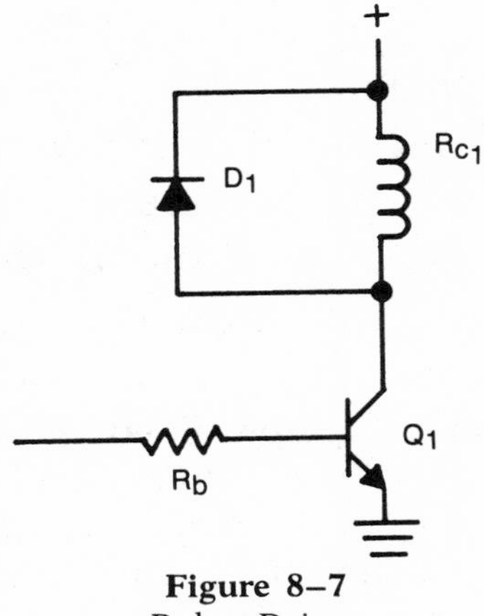

Figure 8–7
Relay Driver

of a relay driver. Before we start calculating values, a comment is in order. D1 is included in this circuit only to protect the transistor from the high counter EMF that is developed in all coil circuits that are switched. It can be any standard switching diode, such as 1N914, or for that matter, any silicon power diode.

Step 1. The load is defined by the relay being used. In this case, the relay coil chosen is rated at 30 ma. at 12 volts. If the relay you wish to use specifies only the coil resistance, use Ohm's law to calculate relay current.

Step 2. Specify output voltage. This voltage should be at least 12 volts since we are using a 12-volt relay. You could use a higher supply voltage than this if desired. Our circuit will operate from a 12-volt source.

Step 3. Calculate collector resistor. If the source voltage is 12 volts, the same as the relay voltage, then no collector dropping resistor is required. Otherwise, subtract the relay voltage from the supply voltage. Use this value, and collector current required to solve for RC, using Ohm's law.

Step 4. The input voltage in this case is from a digital circuit that will deliver a 5-volt positive signal when the relay is to be energized. The input voltage is then 5 volts.

Step 5. Determine transistor type. A look at the specification book, and the parts bin tells me that those 2N2222 transistors will work in this application as well. (This general purpose switching transistor is usable in hundreds of switching applications.)

Step 6. Calculate Ib by the formula. In this case,

$$Ib = 0.6/100 = 0.006 \text{ amps.}$$

Step 7. To calculate Rb, again use Ohm's law.

$$Rb = (5 - 0.6)/0.006 = 733.3 \text{ ohms. Use an 820-ohm resistor.}$$

You have seen how easily a simple transistor switch can be designed. Much more complex switches can be designed for all applications. By some simple calculations, it is also possible to switch a circuit off, when the switch signal goes on, as in the circuit of Figure 8–8. When the transistor is off, current flows through the collector resistor, and through LED 1. Since the transistor is open, there is infinite resistance in the transistor junction, so no current flows through the transistor. To determine the value of the collector resistor, subtract the load voltage (in this case the drop across the LED) from the source voltage. Use Ohm's law to calculate for Rc. Ic is now determined by the value of supply voltage and collector resistor.

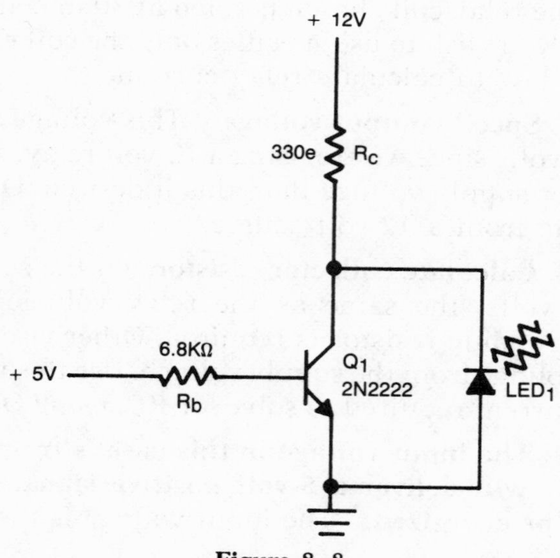

Figure 8–8
Transistor Switch

When the input voltage turns on the transistor, the voltage across the LED drops to zero, and the LED turns off. Calculating for all other component values is exactly the same. Calculate them, and check to see that you obtain the same values. (Remember to use the closest standard value available.)

Switching circuits are not especially critical when determining values, and you will be amazed to find out how you can experiment with the specific values and still allow the switch to saturate. One other point to keep in mind is that all of these formulas are applicable to PNP devices as well; the only difference is the reversal of input and output voltage polarities (and the polarities of any other polarized devices in the circuit, of course).

The experimenter will have a lot of specific applications for these switches. If you become familiar with them, you will find them among the most used transistor circuits in your projects.

How to Make a Transistor Amplify

In the last section, we used knowledge of Ohm's law to design a transistor switch. Since switches are basically DC devices, their actual values are easily calculated, and the design process is simple. In this section, the transistor amplifier is explained and analyzed. However, due to the complex calculations involved in designing amplifiers, actual design is not covered. If you understand the principles involved in operating transistors, you will have all the knowledge you need to repair transistor amplifiers. You will also have an understanding of the operating characteristics of the integrated circuit amplifier. A look at the project section will reveal that IC amplifier design is much easier than transistor amplifier design, and the experimenter who designs circuits today uses ICs whenever possible.

All this does not preclude the requirements of understanding how transistors work. A basic understanding of the different transistor amplifier circuits makes the experimenter's task easier. If you know the function of a given component, for example, it becomes much easier to substitute a suitable value component, should the desired part be unavailable.

For the purpose of study, this section will cover how amplifiers are biased, and how they operate. Class A and Class B amplifiers are discussed separately, and the common emitter amplifier will be discussed exclusively.

A perfect amplifier is a device that takes an AC voltage, makes it larger, and adds no noise or distortion of its own. Of course, the perfect amplifier does not exist, though high-quality circuits approach perfection, at least as far as the ear can tell. Every amplifier circuit can be divided into two parts, the DC bias circuit, and the AC signal circuit. DC bias is required to set up the conduction level of the amplifier, while AC signal is the voltage that is actually being amplified. Let's look at DC bias requirements first.

Figure 8–9 contains a circuit that looks quite familiar. Except for the addition of the two capacitors, it looks quite a bit like a transistor switch. A more subtle difference is in the actual values of Rc and Rb. More on this later. Let's analyze how the circuit works. Rc and Rb bias the transistor so that the device does NOT saturate. In fact, it is usually calculated so that the voltage drop that appears across the transistor is one-half the supply voltage (Vcc). To make things easy, assume Vcc is 10 volts. The voltage across the transistor will then be 5 volts with no AC signal applied. The current drawn by the device with no signal input is called the quiescent, or resting, current. The values of voltage and current at this point are called the DC bias points, or static bias.

As the signal to be amplified is applied to the capacitor, Cin, its voltage adds to and subtracts from the quiescent voltage, thus varying the base current. As you already know, varying the base current will vary the internal resistance of the transistor, and the voltage drop across the transistor will vary in inverse proportion to the applied voltage. The transistor has the ability to amplify because the changes in collector current are much greater than the changes in base current.

If designing amplifiers were as easy as calculating switching parameters, everyone could be a design engineer.

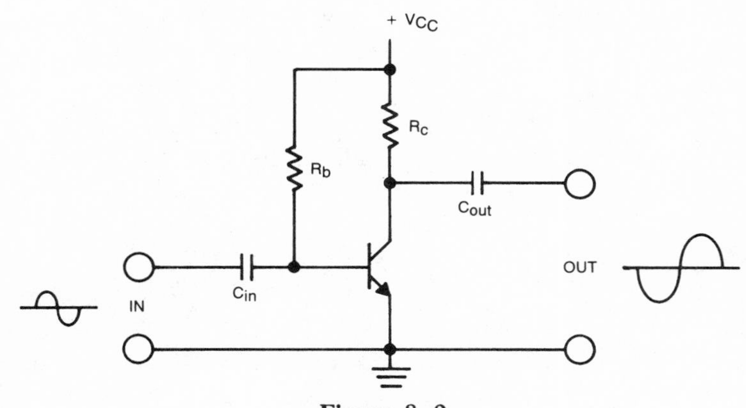

Figure 8–9
Transistor Amplifier

The circuit in Figure 8–9 will work with a given transistor in a given circuit; however, because of individual transistor variations, transistors with slightly different gain (beta) will require a complete recalculation of bias points. Worse than this, the transistor has a stability problem related to device temperature. Due to the nature of semiconductor material, the actual resistance of a transistor will decrease as the device warms up. In other words, when you bias the transistor to amplify, its internal temperature begins to rise because of the collisions of electrons in the device. This increased temperature brings about lowered resistance, and the collector current increases. This increase in collector current causes a corresponding increase in heat generated by the device, which in turn causes an increase in collector current. This nasty little characteristic will eventually destroy the transistor, if collector current ever exceeds the maximum allowed by the device. The term given to this phenomenon is quite descriptive—"thermal runaway." Designers must use special bias circuits that self-correct for thermal runaway, and other changes in the amplifier.

Figure 8–10 is similar to the previous circuit with the exception of the base bias resistor connection. Instead of being connected to Vcc, the resistor is connected to the collector of the transistor. As the transistor warms up, and collector current increases, the voltage drop across the transistor collector to emitter decreases. This results in less

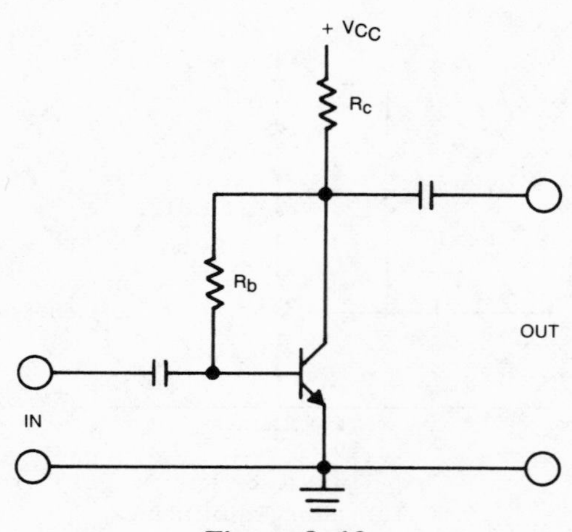

Figure 8–10
Amplifier with Protective Bias

voltage available to power the base circuit. "Self-bias," as this circuit is called, helps to stabilize the transistor by automatically adjusting to the actual demands of the transistor.

Figure 8–11 is similar to the self-bias circuit of 8–10, except for the inclusion of an emitter resistor, Re. Adding the emitter resistor improves the thermal stabilization and decreases distortion. The disadvantage of using the emitter resistor is a decrease in stage gain, and the need for an emitter resistor bypass capacitor (Ceb). This capacitor is chosen for a low reactance to the frequency being amplfied. In other words, the capacitor will allow the emitter to appear grounded to the AC signal. At the same time, a DC bias voltage will appear across the emitter resistor.

Figure 8–12 demonstrates the bias technique used in most amplifiers. Resistor Rs is a stabilizing resistor that works in conjunction with the base bias resistor to further stabilize the transistor. As you can probably see, calculating these values must take a lot of different factors into consideration. As an example, the lower the value of resistance in Rs, the more stable the device becomes. Changing the value of Rs also changes the input impedance of the amplifier, however, and as you will remember from earlier

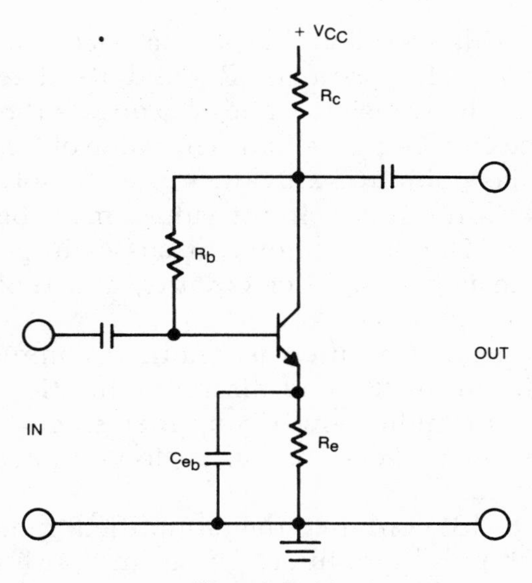

Figure 8–11
Amplifier with Protective Bias

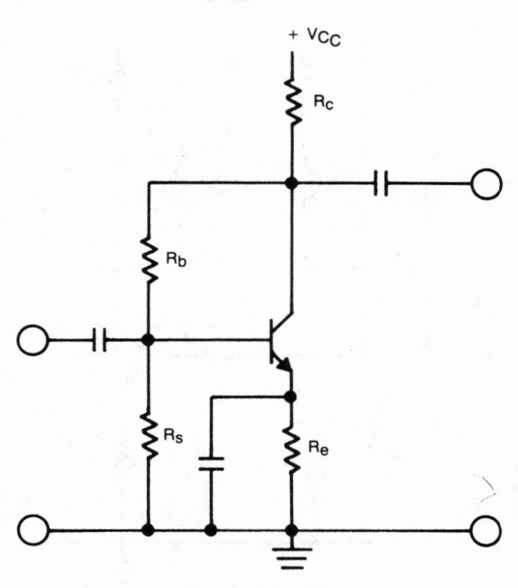

Figure 8–12
Voltage Divider Bias Circuit

reading, impedances should be matched for maximum energy transfer. The ratio of Rb and Rs determines the conduction of the transistor, and also affects the actual bias voltage of the collector terminal. The value of Rc and Re also affects the bias point and all values must be solved simultaneously. In addition, AC circuit values must be taken into consideration. The next chapter discusses in greater detail repair of transistor amplifier circuits, and replacement of circuit components.

The class B amplifier is another commonly found circuit, both within ICs and discrete amplifier designs. Its major use is in audio power amplifier stages, stages that usually drive speakers, or other low-impedance, high-current devices.

Figure 8–13 contains the simplified diagram for a typical "push-pull" amplifier. The term "push-pull" refers to the fact that only one transistor works at a time. First one "pushes" the signal, then the other "pulls" it. Let's see how

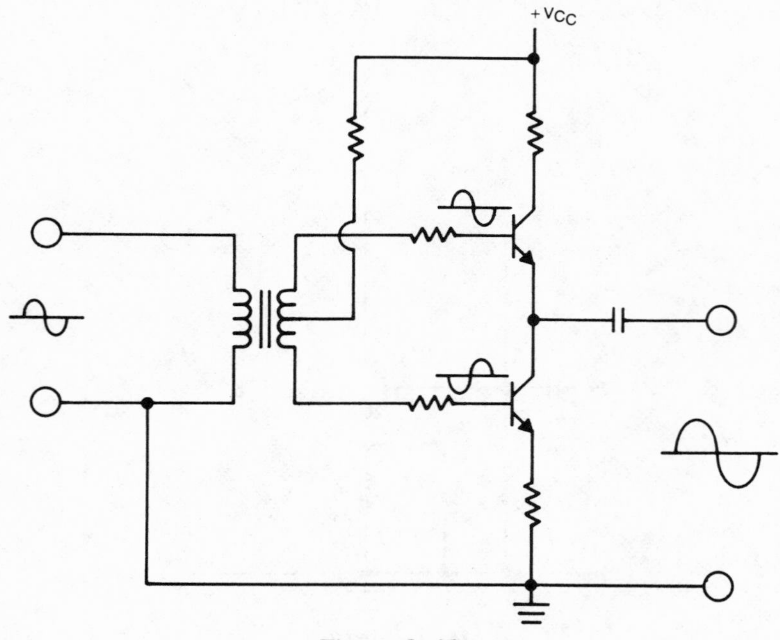

Figure 8–13
Push-Pull Amplifier

it works. Signal is applied to the transformer, and appears at the secondary. Since the secondary is center-tapped, the signals applied to the transistors are 180 degrees out of phase. Base bias resistors are calculated so that quiescent current is zero in both devices (very energy efficient). As the signal goes positive, the transistor with the positive going waveform begins to conduct, increasing the voltage appearing at the output of the circuit. The other transistor is getting a negative going waveform at this time (the two waveforms are out of phase). That transistor is cut-off, and will not conduct. As the incoming waveform goes negative, the situation is reversed, and the first transistor is now cut-off, while the latter transistor is getting the positive going waveform. The lowered resistance of that transistor is reflected by a lowered output voltage at the output of the amplifier.

Figure 8–14 is a simplified diagram of another class B amplifier. This amplifier is called a "complementary sym-

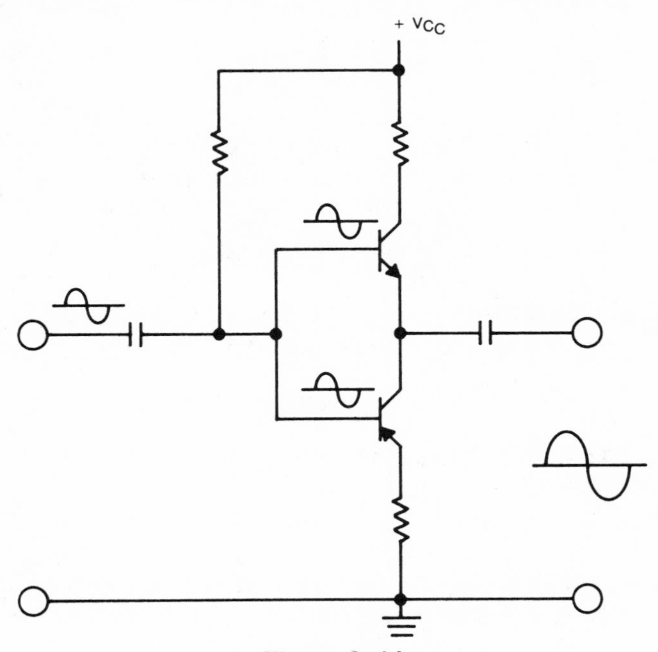

Figure 8–14
Complementary Symmetry Amplifier

metry" amplifier. Simply stated, the name comes from the fact that the circuit uses an NPN and a PNP in its output stage. When the signal is positive going, the NPN transistor receives a positive voltage on its base, decreasing the transistor's internal resistance. At the same time, the PNP transistor is cut off, increasing the internal resistance of that device. The net result is an increase in the output voltage appearing at the junction of the transistors. During the negative-going part of the cycle, the NPN transistor is cut off, while the PNP transistor conducts. The output voltage decreases accordingly. Advantages of the class B amplifier include increased efficiency, and increased power for the same dollar. Since each transistor works only fifty percent of the time, its ratings do not have to be as high as an individual class A device would need.

There are also class C amplifiers, both in single- and double-ended stages (push-pull amplifiers are double-ended). The only difference in operation of a class C amplifier is in actual calculation of bias points. A class C amplifier must, by definition, conduct less than fifty percent of the time. Major application for class C bias is in radio frequency circuits, where the coils and capacitors assist in generating proper sine wave output, even though the amplifier itself doesn't. For more information on class C circuits, consult the *Radio Amateur's Handbook,* published by the American Radio Relay League.

We have covered just about all we need to know about the amplifier for understanding how it operates, and now it's time to look at the oscillator and its functions and applications.

How to Make a Transistor Oscillate

The oscillator is a combination of the switch and amplifier. In other words, an oscillator switches a DC input voltage while the voltage is being amplified. The oscillator may or may not have a single input; its power is obtained from the power supply, as well as the DC that is eventually being switched.

There are three types of oscillators generally in use

today, and we will look at each one separately. The most commonly found oscillator is the LC circuit. A variation on the LC oscillator is the crystal oscillator. The third type to be explained will be the RC circuit. As you will see, all three have their specific applications, advantages, and disadvantages.

The LC oscillator, as the name implies, contains an inductor and capacitor as its frequency determining element. The oscillator, in its action of changing DC to AC, will operate at a certain frequency. The exact frequency of oscillation in an LC circuit is determined by the values of the inductor and capacitor. Specifically, when the inductive reactance of the coil matches the capacitive reactance of the capacitor, the circuit is resonant. In an oscillator, the resonant frequency determines the output oscillator frequency. Before going into the process of describing the action of the active device in the oscillator, it is a good idea to spend a few moments on exactly how a coil and capacitor develop an AC voltage.

Figure 8–15 shows a coil and capacitor tuned circuit. The oscillation process can start with a single pulse of DC voltage. This pulse charges the capacitor in the circuit (Figure 8–15A). Once the capacitor is charged, it tries to discharge through the coil, since a DC path exists (through the coil) to the other capacitor plate (Figure 8–15B). However, as soon as the discharge current begins to flow through the coil, a counter EMF is developed that opposes the change in current. This actually charges the capacitor to the opposite polarity (Figure 8–15C). Since the capacitor is

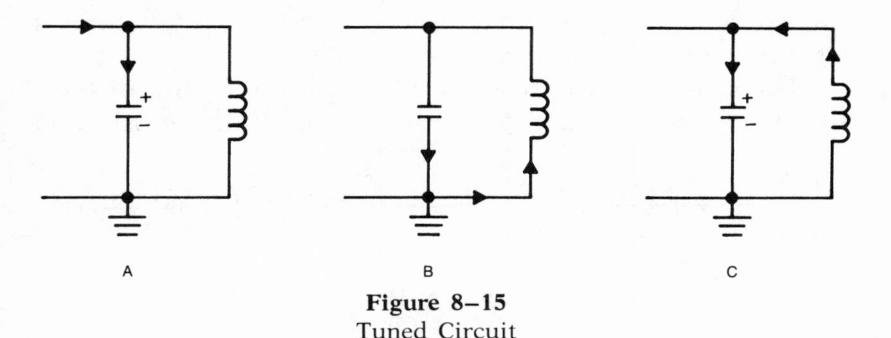

Figure 8–15
Tuned Circuit

again charged, it will again try to discharge through the coil. The cycle repeats itself until there is not enough energy left to sustain the oscillation. This action is called the "flywheel effect" by experimenters. It is also possible to start the flywheel effect by allowing current to flow through the coil, instead of charging the capacitor. When this method is used, the counter EMF charges the capacitor, and the cycle begins.

Figure 8–16 demonstrates the actual output of a flywheel "tank" circuit. "Tank" is a term hobbyists have given to the coil and capacitor. The term is applicable, since it is easily seen that energy from the power supply is stored in the "tank." Figure 8–16A shows the output of a low-Q tank circuit, while 8–16B is the representation of a high-Q circuit. The output of the tank circuit is called a "damped" sine wave. As you can see, the waveform gets smaller and smaller, as the stored energy is dissipated. Q, you will recall, is a measure of the quality of a resonant circuit. High-Q circuits have a high reactance compared to resistance ratio. The greater the resistance in a tank circuit, the lower the circuit Q.

All that is left to do to make this circuit into a full-fledged oscillator is to provide a pulse at the proper time. This pulse will recharge the capacitor, so that the energy lost in the tank circuit can be replenished. By putting an amplifier at the output of the tank circuit, the small damped waveform is amplified. A portion of this amplified signal is routed back to the tuned circuit to recharge the capacitor. Experimenters call this process positive feedback. Since only a sample of the pulse is returned to the tuned circuit, the AC generated by the oscillator, and amplified by the transistor, can be used by the experimenter for whatever purpose he or she desires.

There are several types of LC oscillators, and all basically work on the same principle. Probably the most com-

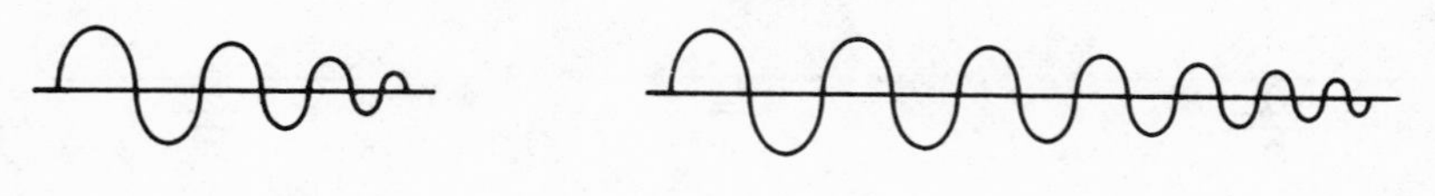

Figure 8–16
Tuned Tank Circuit

monly found oscillator is the Hartley. Figure 8–17 is a typical Hartley oscillator. Let's analyze it and see how it works. Energy is applied to the tank circuit via the battery. The flywheel action begins, sine waves being amplified by transistor Q. Notice that the tank circuit capacitor is variable, which allows the experimenter to vary the output frequency of the circuit. R1 is a bias resistor to allow the transistor to operate. The feedback path is through C3, the amplified sine wave being applied to the lower side of the coil. The positive feedback sustains the oscillations generated. The output of the oscillator is coupled to an external circuit through L2. L3 is included in the battery line, and is called an RF choke. The coil is calculated to have a high reactance to the oscillating frequency, and prevents the AC from getting into the battery circuit and being lost.

Other LC oscillators have different names; for example, the Colpitts and the Armstrong, among others. The principles of operation are quite similar and differ only in minor circuit details. All have a tank circuit, an active amplifying device, and a feedback path. If you are interested in exploring the operation of these oscillators, a good reference source

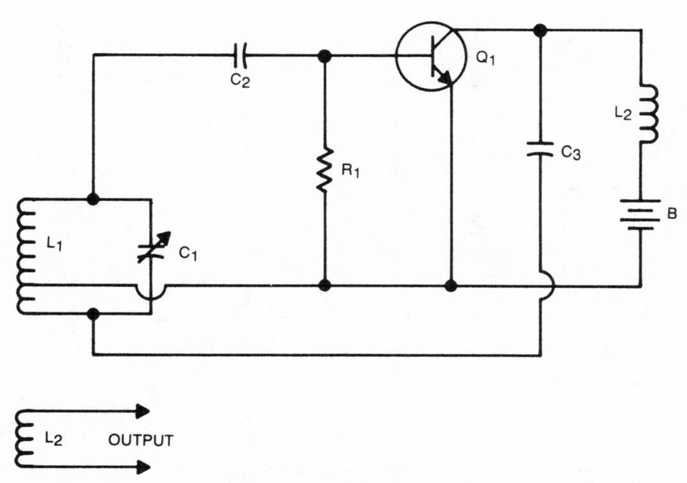

From the book, Complete Guide to Reading Schematic Diagrams, *2nd edition, by John Douglas-Young © 1979 by Parker Publishing Company, Inc. Published by Parker Publishing Company, Inc., West Nyack, New York.*

Figure 8–17
Hartley Oscillator

is the *Complete Guide to Reading Schematic Diagrams,* second edition, by John Douglas-Young; Parker Publishing Co., 1979.

The crystal oscillator is commonly found in circuits that require high-frequency stability. In ordinary LC circuits, minor power supply voltage changes, temperature changes in the transistor, and even component placement affect the actual ouput frequency. In many cases, this is not a problem. However, if a high percentage of output frequency accuracy is required, a more stable frequency determining element is required. Experimenters have called upon the crystal to provide this accuracy. Remember, a crystal, when excited, produces an output waveform that is quite stable in frequency and amplitude. Figure 8–18 is a Colpitts crystal oscillator. Essentially, the crystal, Y, replaces the tank circuit. C1 and C2 are required to ensure a positive feedback path, and if one or the other is made variable, the actual crystal frequency can be shifted slightly from its nominal value. Resistors R1 and R2 are voltage divider bias resistors

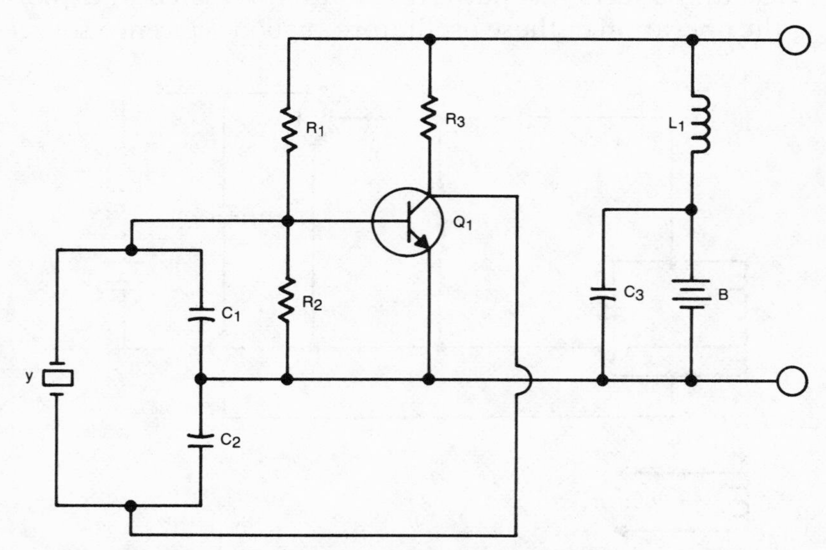

From the book, Complete Guide to Reading Schematic Diagrams, *2nd edition, by John Douglas-Young © 1979 by Parker Publishing Company, Inc. Published by Parker Publishing Company, Inc., West Nyack, New York.*

Figure 8–18
Colpitts Oscillator

for the base lead, and R3 is the collector supply resistor. Feedback is obtained by connecting the collector to the junction of the crystal and C2. L1 is an RF choke, and C3 routes any AC getting through the choke away from the battery and back into the amplifier circuit.

The last group of oscillators commonly in use is the RC type. The RC oscillator contains no coils, and instead of generating sine waves, it generates pulses, or square waves. Also, unlike LC circuits, which can operate at frequencies well into the VHF range and higher, and crystal circuits, which operate at several megahertz, the RC oscillator is not used much above the audio frequencies. Figure 8–19 is a two-transistor multivibrator circuit. While this seems like a much more complex circuit than the others we have analyzed, it really is not so. Q1 and Q2 alternately conduct and

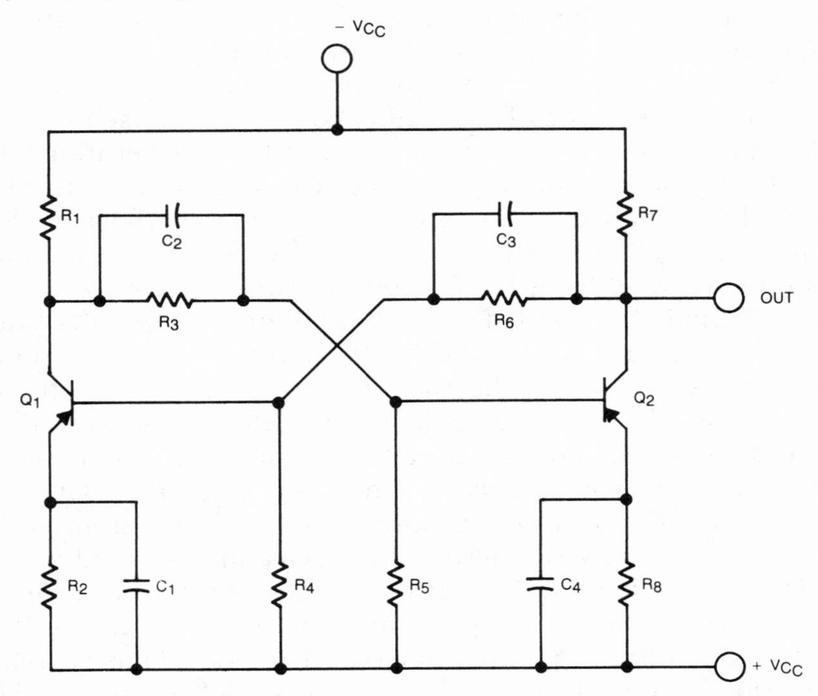

From the book, Complete Guide to Reading Schematic Diagrams, *2nd edition, by John Douglas-Young © 1979 by Parker Publishing Company, Inc. Published by Parker Publishing Company, Inc., West Nyack, New York.*

Figure 8–19
Multivibrator Oscillator

cut-off, providing a square wave pulse between +Vcc and OUT. The frequency of oscillation is determined by C2 and C3 in conjunction with the bias resistors. Bias for Q1 is established via R4, R6, and R7, while Q2 bias is obtained from R1, R3, and R5. C1 and C4 are emitter bypass capacitors to allow AC to return to +Vcc. R2 and R8 are emitter resistors. An interesting aspect of the multivibrator is that the duty cycle can be varied. The duty cycle is the ratio of on time of Q1 compared to Q2. If C2 and C3 are equal (assuming equal bias resistors), both transistors are on as long as they are off, and the duty cycle of each transistor is fifty percent. Changing the value of one or the other capacitor will change the duty cycle, allowing one transistor to conduct for a longer period of time than the other. This feature allows us to change the pulse durations from short negative pulses, far apart, to short positive pulses, far apart. Proper selection of component values can give the experimenter the duration and frequency of pulses required.

The three types of oscillators commonly in use provide the experimenter with much versatility in generating AC voltages for any number of applications. Oscillators are found in just about every consumer electronics device made. In Chapter 11, you will construct an IC oscillator to generate an audio wave form for test purposes.

It should be pointed out that oscillators, amplifiers, and switches do not have to be made from bipolar transistors. For years and years, the vacuum tube was used for these functions. Though not completely obsolete, the vacuum tube has definitely fallen upon hard times, and its days as a useful electronic device are certainly numbered. Another device I have not mentioned before now is the field-effect transistor, or FET. You may run across this device in your reading or circuit construction. The basic difference between a field-effect transistor and a bipolar junction transistor is that the FET is a voltage operated device, while the BJT is current operated. Amplifiers, oscillators, and switches can also be easily constructed using FETs. You will find that FETs are just as easy to use as transistors. There is a whole range of semiconductor devices available today that could be discussed, including such examples as the unijunction transistor (UJT), the silicon controlled rectifier (SCR), the Gunn diode, and many others. All of these devices can be used in at least one of the three functions of amplification, oscillation, and switching. If you are interested in solid-state electronics, you can specialize in just about any circuitry and components you could imagine.

Now that you know about the basic building blocks of electronic circuits, we can spend a little time learning how to tell when a circuit doesn't work, and exactly what to do when it doesn't. Chapter 9 covers troubleshooting procedures, and explains the application and uses of the block diagram. You will learn to use the signal tracer you build in Chapter 11 to locate problems in a circuit. Also included are troubleshooting charts that will assist you in locating problems in several different kinds of equipment.

How To Isolate Problems In Electronic Equipment

9

This chapter puts together everything you have learned up to this point. By now, you should have an understanding of basic electronic concepts, and a good background in solid-state electronics. You should also have a feel for working with the three building blocks of electronic circuits—the switch, amplifier, and oscillator. Let's put this knowledge together to repair defective electronic circuits.

How to Use the Four-Step Troubleshooting Method

Given any piece of electronic equipment, the experimenter can determine the extent of a problem and, in many cases, repair it, without a schematic. All you need is a logical approach to troubleshooting, and a reasonable block diagram. I don't want to imply that any electronic device can be repaired without access to a schematic, as complex circuits can get very difficult to follow without a "road map."

There are four logical steps that can be applied to the repair of any electronic equipment. All troubleshooting follows this process, whether the experimenter knows it or not. Troubleshooting is basically, first, removing from consideration all components that cannot possibly cause the problem, then, from the components that are still suspect, using basic electronic theory to determine the defective components.

The first step in troubleshooting is to identify the symptom. This seems obvious, but it can be more complex than it first appears to be. Complex equipment requires more time on this step than simple equipment. For example, a tape player has no sound in the left channel. Before even opening the case, the experimenter realizes that the left channel amplifier is defective. A television has no picture. This is a much more complex problem. First, is the screen light or dark? If dark, is there high voltage? If yes, is there a video or AGC problem? There are a lot of circuits that can cause loss of picture on a TV set, while there are only a couple of circuits that will cause no output in the left channel. The last section of this chapter has several flow charts that will help you to identify an exact symptom. Sometimes symptoms can be deceiving. In some cases, what looks like a simple problem with an easily identified symptom turns out to be something else. More detail on symptom identification and the flow charts as a troubleshooting tool, later.

Once you have determined the symptom, you will have a good idea as to which block is defective. Let's back up a second. What is a block anyway? In its simplest form, a block is a single stage in a piece of electronic equipment. A single amplifier, oscillator, or switching device and its associated bias components are a block. On a larger scale, a block may be a small group of several active components and their associated circuits. An example of this might be a power amplifier block in a stereo. This block may contain two or three or more active devices. More details and some concrete examples are listed in the next section. Ideally, the experimenter would like to trace a problem to the smallest possible block, as this gives him or her fewer components to check.

After determining which block is defective, the experimenter must look at the block, and, using the principles of electronics learned, determine which component or components in the block are defective. In our example of the defective left channel of a stereo, you would apply what you learned from the last chapters to check the components for obvious defects—proper supply voltage, or proper signal input. The final objective is, of course, to determine which part or parts must be replaced.

The last of the four steps is repair and test. This is probably the simplest of the steps, though sometimes it is the point at which the experimenter must stop. It is a shame when a technician has spent a large amount of time and found the defective component, then tries to obtain a new one, only to find it unavailable. The component in question may be unmarked, and you have no part number or other data with which to order a replacement. It is possible to use your knowledge of circuitry, and, with a little help from this book and maybe a reference guide, to determine enough about a device to allow you to select an adequate replacement.

Once you have completed repair of the unit, a test period, or burn-in follows. This is to ensure that the problem is indeed corrected, and that there are no other problems that intermittently appear. Another thing to do at this point is to realign and readjust any adjustments that were affected by the installation of the new component(s). All controls and contacts are cleaned, boards are checked for poor solder connections, and a general cleanup is in order.

Now that you have an overview of the process, we will look at each step in greater detail. Symptom identification and the block diagram are first. We will look at the use of a signal tracer to assist in locating the defective block. This device is essentially the culmination of several projects that are built in Chapter 11. Circuit and component testing comes next, to locate the problem. We will look at replacing the defective component last, then you will find several flow charts that will assist you in locating defective blocks in several different kinds of electronic equipment.

The Block Diagram and How to Use It

In the last section, we touched upon the block diagram. As a result, you probably already have a good idea as to its use. Let's take a closer look at diagrams for equipment you might use. Figure 9–1 is the block diagram of a simplified audio amplifier. The input of a block diagram is on the left and the output is on the right, just as it is in a schematic diagram. As you can tell by the direction of the arrows, the signal enters into the preamplifier. It is transferred to the driver stage, then is sent to the power amplifier. The audio is then delivered to the speaker. Nearly all circuits require a power supply, and one is shown in the diagram. Many diagrams, however, do not show the power supply block, since it is understood that the block is required for proper operation of the circuit. Unless specifically necessary, this chapter will follow that convention, so you must assume that each circuit needs, and has, a power supply block.

It was said earlier that the diagram shown is simplified. Only a low-power amplifier would have just three stages. Also note that there are no volume or tone controls in this diagram. You might wonder why we would not always use a complete block diagram. It is often not necessary to have a complete diagram to understand the operation of a typical unit, therefore a simplified diagram is adequate. We will use both in this chapter. Let's look at how a complete block diagram of this unit might look. Figure 9–2 contains a much more detailed diagram of our hypothetical amplifier. Notice that the preamp is actually two transistors. Q1 is designed to take the very weak signals from a tape head and magnetic phono cartridge, and amplify them. The output of Q1 is too weak to deliver proper voltage to the driver stage, so Q2 amplifies the signal further. The preamp block in

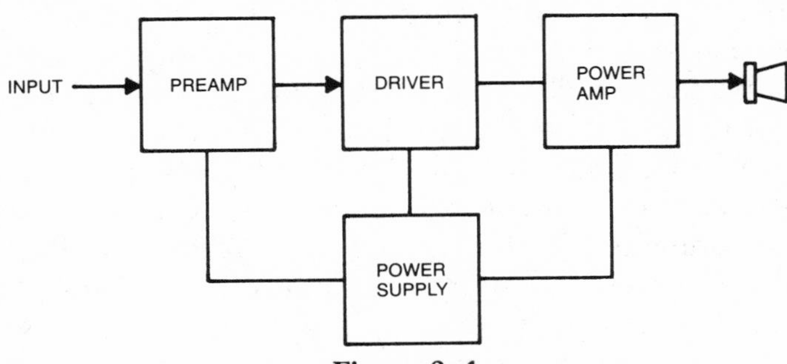

Figure 9–1
Amplifier Block Diagram

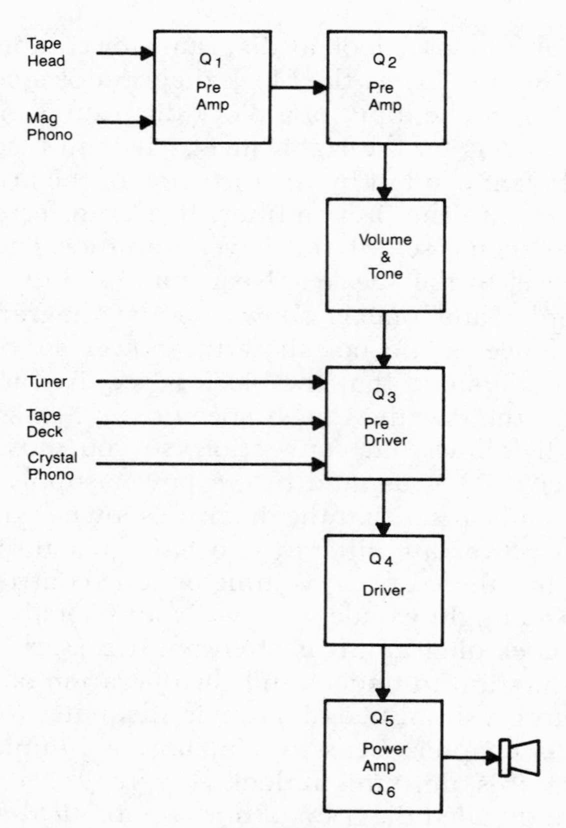

Figure 9–2
Detailed Block Diagram

Figure 9–1 has two transistors in it. Each active device that is in our amplifier is shown in this block diagram. The DC bias and power supply components are not shown.

Moving on to the volume and tone block—these components are passive and need no power supply connections, and the projects in Chapter 10 will show the actual circuits. Tone controls are basically passive filters, either variable high pass, for treble controls, or variable low pass for bass controls. Volume controls are usually just variable resistors that are connected in the signal path. When you adjust the control, you are changing the amount of voltage that is leaving the stage. Notice that high-level output devices, such as tuners, tape decks and ceramic phono cartridges input at this stage.

After leaving the tone and volume block, the signal is amplified by Q3, the predriver. The driver transistor Q4 provides further amplification. As with the preamp block, there are two active devices represented by one block in Figure 9–1. After leaving the Q4, the signal is delivered to the power amplifier, which is a push-pull stage composed of Q5 and Q6. Push-pull stages are often represented by one block, even though they contain more than one transistor. They are sometimes also shown as two blocks.

The actual number of active devices, and therefore the number of blocks, will vary from one device to another. The greater the power output of an amplifier, the more stages of amplification are required.

Figure 9–3 contains the block diagram of a typical AM radio. It has a few new stages that we can look at. As you can see, it is a little more complex in its terminology; however,

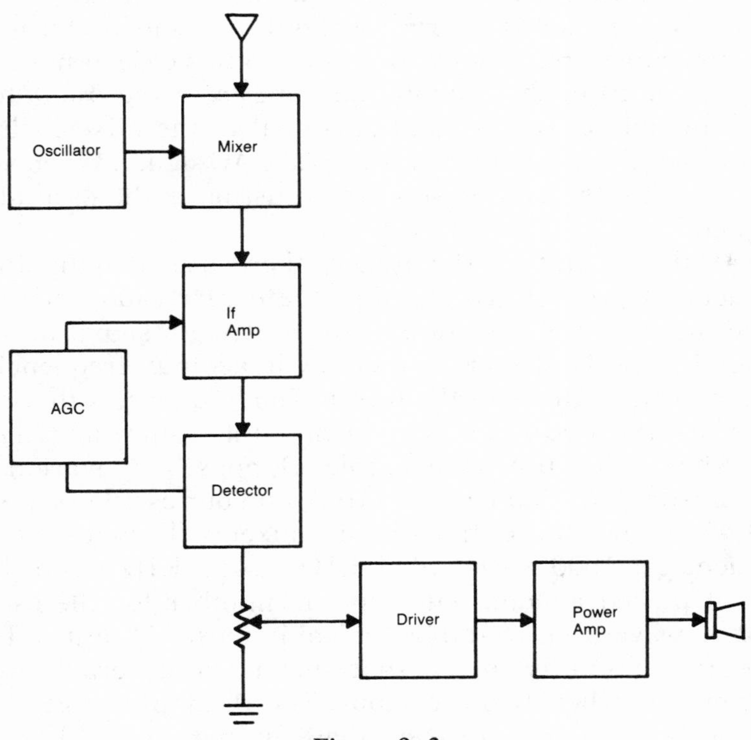

Figure 9–3
AM Radio Block Diagram

there are actually only six active devices in this particular radio. Notice also, the potentiometer between the detector and driver block. This represents the volume control. Many block diagrams will insert a schematic symbol or two for passive devices. The speaker and antenna symbols also happen to be included in this diagram. Let's look at the function of each block, and see what each one does. As with all block diagrams, let's start at the left.

The purpose of a radio receiver is to obtain the signal from the station, and process it to remove the audio component. This audio component is amplified, and sent on to the speaker. The radio frequency itself is discarded by the receiver in the detector circuit. As we proceed through each block, we will look at the process in more detail.

The antenna feeds its signals into a block called the mixer. This is a new term; however, as you already know, there are only three basic circuits—amplifier, oscillator, and switch. The mixer is a form of amplifier, and performs a function called frequency conversion. In this case, it has two inputs, one from the antenna and one from the oscillator. The antenna delivers all station signal to the mixer, while the local oscillator delivers a sine wave AC signal. When you tune a radio, the circuit you are adjusting is the oscillator frequency.

At the output of the mixer, there are actually four frequencies present. To simplify the explanation, let's assume there is only one station transmitting a signal to this radio. If this is the case, there will be four frequencies present at the output of the mixer. The frequencies that are there are the two input frequencies, their sum, and their difference. Let's use an example. Suppose our station is transmitting on 1000 KHz. (1 MHz), and our oscillator is set to 1,455 KHz. The output of the mixer will contain four frequencies: 1000 Khz.; 1455 KHz.; 2455 KHz.; and 455 KHz. I picked an unusual-sounding number for the oscillator, however that would be typical in most AM radios. The important point is to see that the mixer generates four frequencies, when two are input. The IF amplifier gets all four of these signals; however, there are band-pass filters at the input and output of this amplifier that reject all but the 455 KHz. signal. IF stands for "intermediate frequency," and is named because the IF is between the desired received frequency and the actual audio signal we are obtaining from the station. More on this later.

There are many more signals than one entering the antenna, but the selectivity of the input tuned circuits rejects all but the desired frequency. As a result, there are four frequencies present at the output of the mixer—the station signal, the local oscillator, and the sum and difference frequencies. The IF amplifier will reject all undesired frequencies. By tuning the oscillator, you are able to allow another station to create a 455 KHz. signal for the IF amplifier. As an example, if we change the oscillator to 1655 KHz., the signal that produces a 455 KHz. difference frequency is 1200 KHz. Tuning the oscillator will assure that all desired frequencies will be converted to 455 KHz., and be amplified by the IF amplifier.

Many radios contain only a single device that acts as both oscillator and mixer. When this is done, the block diagram labels the stage as a convertor. As you can see, the function is to change all station frequencies to one frequency called the IF amplifier. This seems like a complex way of receiving radio signals; however, it has two definite advantages. The high gain and selectivity of the IF amplifier and the stability of the local oscillator are the chief reasons that this mixing principle is used in nearly all communications reception, from radio and television to satellite data transmissions.

As was stated earlier, the IF amplifier has band-pass filters that reject all frequencies except the IF frequency. At this point, it is possible to amplify the IF frequency to any desired level. The AM radio usually has only one stage of IF amplifier, however the block diagram of an FM radio could have three or four stages of IF amplifier. Televisions have as many as five stages, and communications equipment could have even more than that.

If you wish to receive a certain station, and drive a speaker to a high level, many stages of amplification are required. Remember, signals on the antenna are measured in microvolts. We need a couple of volts to properly drive our detector (more on this later), so high amplification is in order. Tuning each stage to the station frequency is not practical. Every time you wanted to change stations, you would have to retune each amplifier for the new frequency. By using a convertor, or mixer-oscillator, you can tune the IF amplifiers only once when the radio is manufactured. The only time these adjustments must be changed is after repair, or occasionally as components age. The process of retuning these amplifiers is known by experimenters as the alignment procedure. On AM radios, the process is simple; FM and television receivers have a much more complex alignment process. There are other common IF frequencies, besides 455 KHz. The AM automobile radio uses 262.5 KHz., FM radios usually use 10.7 MHz., and televisions use 45 MHz.

The intermediate frequency amplifier is the heart of this process. By using high-Q band-pass filters, and many stages of amplification, you can deliver much power from a minute radio signal. It is not at all unusual for a high-quality communications receiver to deliver four or five watts of audio from a station that is delivering only 0.1 microvolts to the antenna. The process of frequency conversion and IF amplification in receivers is known to experimenters as the "superheterodyne" process. Heterodyne means the mixing of two frequencies.

After being converted and amplified, the radio signal is delivered to the detector. The detector is often only a half wave rectifier and filter assembly in AM radios, however it is more complex in FM and communications receivers. For our purposes, the signal leaving the detector is the recovered audio, sent by the station. The filter removes the RF component, and runs it to ground. If you desire more details on the detection process, the *Radio Amateur's Handbook* published by the American Radio Relay League is a good source.

After leaving the detector, the signal, now audio, is sent to the volume control. This is the point at which a component tuner usually ends. The output of the tuner is sent to the audio amplifier. In a radio, that is also the case. The driver, in small radios, can be only a single device, or more, depending upon the power requirements of the designer. As in the amplifier circuit discussed previously, the signal is sent to the power amplifier, and then to the speaker.

Now that you know about the operation of the radio and the amplifier, we can discuss how to use the block diagram in troubleshooting. Since each stage has a particular function, an input and an output, we can easily check the performance. We can use a device I like to call a "sumthin probe" to check out each block. The probe might be hooked to an oscilloscope to get a visual picture of the stage condition. However, we will use the sumthin probe on a device called a signal tracer. For now, think of a tracer as a block with a red light on it. The next section will go into more detailed operation of the device.

The red light on the tracer indicates a fault condition, in our simplified analysis. For example, we locate the signal at the output of the mixer. If it is present, and correct, the mixer circuit is working, and our red light stays off. We then place the sumthin probe at the output of the IF amplifier; again, if the signal is of the proper amplitude and frequency, the red light stays off. Assume that the detector circuit is faulty—the signal into the detector is correct, but the signal at the output is missing. The red light goes on. Now you must use your skills and understanding of electronic principles to check the components in the detector block, and effect the repair.

The output of each block could be one of four states. They are: normal signal, missing signal, weak signal, and distorted signal. It is possible to have a combination of these—for example, weak and distorted. The process of localizing the defect to a specific block is sometimes easy if the circuit is simple, and schematics and other service literature are available. It gets more difficult as the device increases in complexity, decreases in physical size, or lacks sufficient documentation. We will look at these problems in greater detail as we continue.

The red light that is on that signal tracer actually is in your thought processes only. You must listen to the output of the tracer to determine if the signal is correct, and if not, the light should go on. Learning this skill is difficult unless you have adequate practice. The easiest way to get practice is to use an amplifier or radio and the tracer you complete in the next chapter to practice on. The next section shows you the proper procedures, and takes you, step by step, through the tracing process.

Using a Signal Tracer to Isolate Problems to the Defective Block

The signal tracer you will build in Chapter 10 is actually the back end of a radio. Its block diagram is shown in Figure 9–4. The device has two inputs, an RF input and an audio input. The switch selects which input is used. When tracing in RF and IF circuits, you must use the detector input. Audio circuits use the audio input.

To use the tracer, connect your sumthin probe to the tracer and turn the gain control of the tracer to maximum. We will troubleshoot a radio that has no sound output. We have already determined that the power supply is functioning. Remember, if you are actually going to use these procedures on a transistor table radio, be sure you use proper safety procedures, and don't forget the isolation transformer. Better yet, use a battery-operated receiver. All tests done using a signal tracer are dynamic; that is, power is applied to the circuit being tested. Figure 9–5 contains a simplified schematic diagram of a radio to help you associate the individual components with their blocks. If you are trying these procedures, read through them first, to get

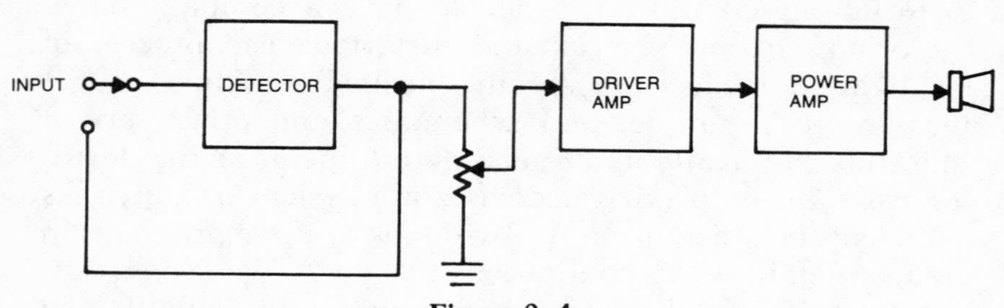

Figure 9–4
Signal Tracer Block Diagram

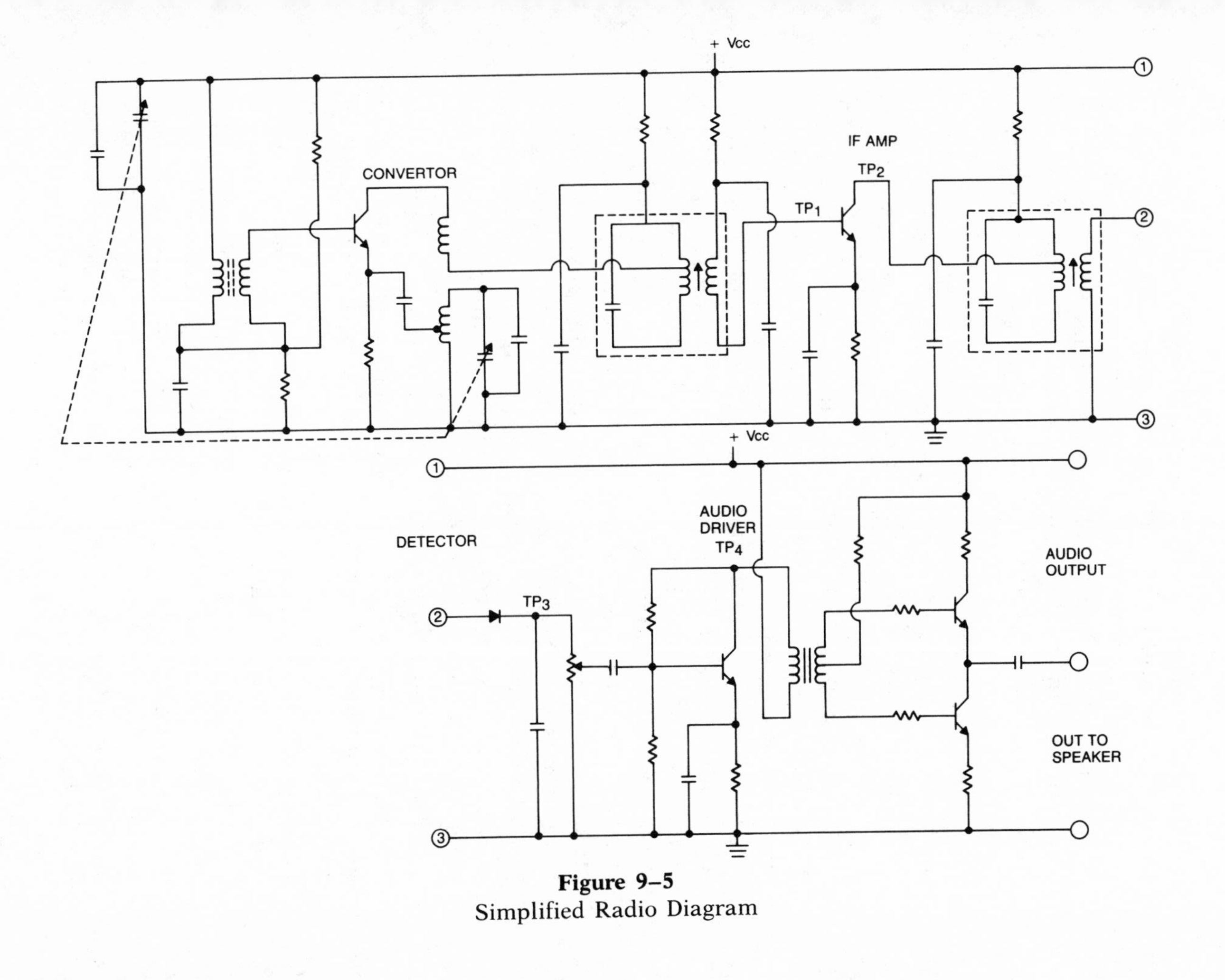

Figure 9–5
Simplified Radio Diagram

an understanding of the process. Then try to obtain the actual schematic for the radio you are using. This can be ordered from the manufacturer, or in some cases, through Sams Photofacts, a private schematic diagram service. Sams has a master index that you could obtain through a local TV-radio shop. If your radio is listed in there, you can order the schematic.

Place the probe at the input of the IF amplifier stage, test point 1, and tune the radio, trying to find a local station. The mixer stage is not tested directly, as it is good to have the first band-pass filter of the IF stage in the circuit. Any signal that appears there will be quite weak, due to the fact that little amplification has occurred to this point. If there are no local stations in your area, you may not hear a signal at all. For now, assume that you hear a signal at this point. This indicates that the mixer stage is working correctly. If there were no signal here, we could have a problem in the mixer circuit, or it could be just too weak to operate the tracer. If this happens, move on to test point 2. A missing signal at that point means that either the mixer or the IF amplifier is defective. Troubleshoot each stage, looking for proper bias and circuit operation. More details on this in the next section. As you can see, there is not always a cut-and-dried test as to which stage is defective.

Move the sumthin probe to the output of the IF amplifier marked test point 2. The signal here should be louder than at the previous signal point, since there is another stage of amplification. If you had a signal at test point 1, and none at 2, the problem is obviously between these two points. The AGC circuit may be causing the IF amplifier to be inoperative, or the stage itself could be bad. Again, use the information in the next section to troubleshoot the stage.

Test point 3 is at the output of the detector. Before attaching the sumthin probe, put the detector switch into the audio position. If the detector is working properly, you should hear reception in the signal tracer, as you tune the main tuning dial. No signal here indicates a problem in the detector circuit.

Test point 4 is at the output of the audio driver. A louder signal should appear here, due to the amplification of the driver transistor. Last, test point 5 is at the output of the power amplifier. No output here indicates a faulty power amplifier. If you obtain a signal at the output of this stage, the speaker must be defective.

As you can see, the process is easy. It is harder, however, to implement on some units than others. Units that are made for serviceability will have printed circuit boards marked with component references, easy access to components, and easy removal of boards, etc. If you spend any time working with electronic equipment, you will have no trouble determining which units are easily serviceable.

This chapter is not meant to turn you into a seasoned troubleshooter. It should help you to understand signal flow in electronic circuits, and assist you in repairing projects that don't work properly. When you construct projects, you will soon learn how to make them serviceable. If you don't, you will be the victim of your own layout, when the time comes to troubleshoot your project.

Now that you have determined the defective block, you must find which component in that block is defective. Let's see how it is done.

Isolating the Problem to the Defective Component

You have determined which block is defective, and now must determine what is wrong with it. This step can be easy or hard, depending upon several variables. There are four steps, however, that will bring the problem to a successful solution in most cases. These steps are:

A. Check the stage for obvious defects. Look for burned resistors, feel and smell for overheated components. Check for bad solder connections, broken wires, etc. Use your senses to determine if the fault can be spotted. The novice experimenter quite often does not spend enough time on this stage, and moves on, then finds later, in many cases, that taking the time to do this step properly would have saved time in the long run.

B. Check bias on active components. Using your voltmeter, look for 0.6 volts difference between base and emitter on a silicon transistor (0.2 for germanium). Look for the proper collector voltage. Is it enough? Is it the correct polarity? Practice, and your knowledge of transistor operation, should allow you to make a reasonable attempt at determining if the stage is working properly. If bias is not proper, examine why not. Open power supply connection? Open bias resistor? Bad transistor? Use your knowledge of how these components work to determine the problem.

C. Check the active device. This goes for tubes as well as transistors. In fact, if you are working on a tube unit or a transistor unit that has socketed transistors, you will probably want to perform this step before step B. Remove the transistor from its circuit and use an ohmmeter to determine whether the transistor is good. Integrated circuits are easy in this respect. If you have power to the device, and signal on the input, a simple check at the output will determine the status of the device. You should have already tested to see if there was output when you determined that the block was defective. No output from an IC with power and input present indicates a defective IC. Further testing is difficult; substitute the IC with a known good component.

D. Check the passive devices. If all of the previous steps have failed to determine the problem, test the passive devices in the block. Check coils, resistors, and capacitors, in that order, for opens and shorts.

If all of these tests indicate negative, you have probably misdiagnosed the block. Return to the block isolation process and double-check your diagnosis.

In the next few paragraphs, we will test the performance of an audio amplifier stage. We have already determined that there is a signal at the input (base), with no signal at the output (collector). Figure 9–6 is the schematic of our test circuit. If you have a schematic showing the voltages that should be at the three transistor terminals, you have an advantage. We will assume we don't have that information.

A visual check of the circuit shows no observable problems. Measuring the collector, we find that it is +6 volts. Your knowledge of transistor operation should tell you

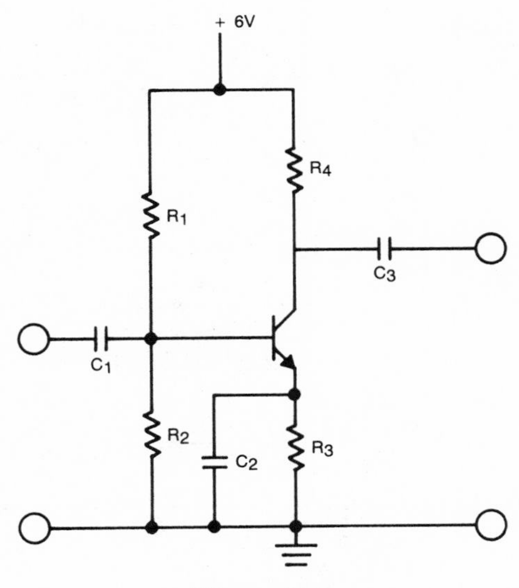

Figure 9–6
Test Circuit Schematic

that the transistor is cut off. Measure the base-to-emitter voltage. If it measures 0.6 volts, the transistor collector is probably open, and you can go on to step C and check it. If the base-emitter voltage measures more than 0.6 volts, the base-emitter junction is probably open. If bias is less than 0.6 volts, either the transistor is shorted, or a bias resistor is open. If the emitter voltage is at or near 6 volts, the emitter resistor, R3, could be open. If it is near 0 volts, check the bias resistors, R1 and R2, for problems.

As you can see, the problem is to look at the operation of the suspected device, test it against the laws and rules you know to apply to the device and its associated components, and draw your conclusions. Practice will bring proficiency in applying the procedures and diagnosing the fault.

Figure 9–7 is a diagram of a transistor circuit we haven't talked about before. It is found often enough, and I thought it would be a good idea to explain the procedure for troubleshooting this special case. The circuit contains two

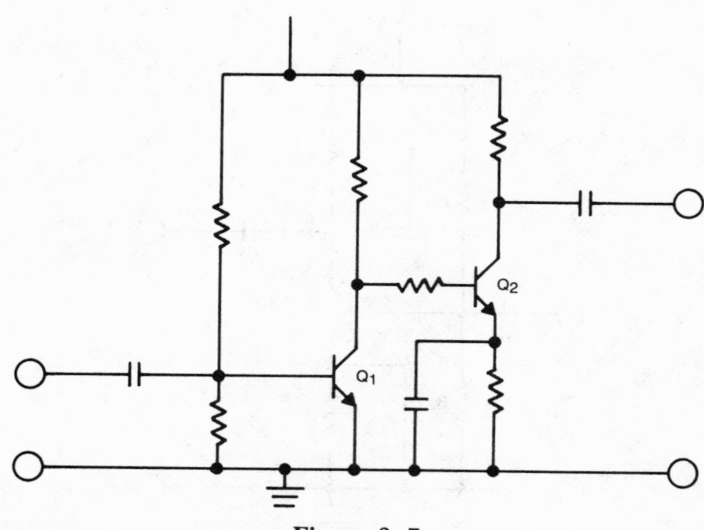

Figure 9–7
DC Transistor Circuit

transistors that are "direct coupled." Direct-coupled transistors have no capacitors or transformers to isolate the output of the first stage from the input of the second. Many amplifiers use this technique, as it improves frequency response of the amplifier, and it is usually less expensive, as coils and capacitors are relatively expensive. Treat these direct-coupled devices as a single stage. This will make it a little easier to troubleshoot. If you run into circuits containing more than two direct-coupled devices, you could have a major diagnosis problem on your hands. Any fault in the bias of Q1 could be reflected in the bias of Q2. As you can see, the collector voltage of Q1 is the bias source for Q2. If Q1 collector voltage is incorrect, there will definitely be problems in testing for the proper operation of Q2. Check each device individually, replacing any you find defective. If the circuit still does not function, start testing components in the stage, starting from the left. Treating these direct coupled devices as a single stage is probably the most efficient troubleshooting method. Don't just replace the first transistor that tests defective and try it again. The new transistor may be destroyed by improper operation of the rest of the circuit. Check **ALL** active devices first, then change any defective parts.

Here are some further hints for testing various types of equipment.

Stereo amplifiers are a definite advantage, when it comes to ease of troubleshooting. Since a stereo has two indentical amplifiers, you can compare the operation of the good channel to that of the bad channel. You can compare signal tracer output levels, DC bias voltages, etc. If you need to, you can even substitute suspected components from the good channel into the defective channel, and verify your diagnosis. If both channels are dead, you are given the clue that the power supply is probably defective. Chances are that both amplifiers won't fail at the same time.

Integrated circuit audio amplifiers have a single output pin in most cases. Figure 9–8 is a representation of a typical audio output IC. It has a few passive components associated with it, and gets its power from a 12-volt source. You have determined that the 12-volt source is present, and there is a signal at the input pin 1. No signal is heard with the tracer, at pin 4. The output pin of these devices should be approxi-

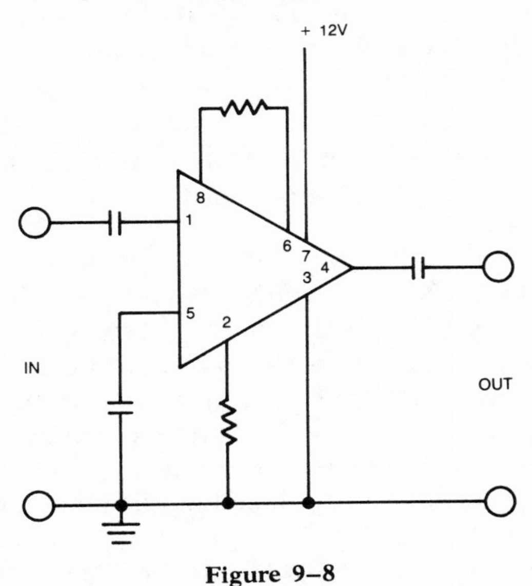

Figure 9–8
Audio Output IC

mately one-half the supply voltage, and relatively stable. Any large fluctuations in DC voltage here, or any voltage more than 20% higher or lower than 6 volts, indicates a fault in the IC. Substitute a known good part.

Troubleshooting switches is easy. If the switch is defective, it will either be open or closed all the time, or it will be conducting, but not saturated. Any of these symptoms is easy to spot. If the switch output is always zero volts, (switch is on), take a clip lead and connect the base to emitter. This will immediately shut off the device. If the collector voltage does not rise to the off state, the transistor is bad. If it does, check the external bias components, because something is not allowing the transistor to shut off.

Oscillators are difficult to test without an oscilloscope. Measure the voltages on the device; again bias must be proper for the circuit to function. Test the active device using an ohmmeter. If all other tests do not point out any faults, and if there is a crystal in the circuit, you will probably want to substitute it.

Now that you have determined the defective part, it is time to order a replacement. This can be a difficult job if proper documentation is not available. The next section gives some hints on finding an adequate substitute for a particular component.

How to Select an Adequate Replacement Component

Obtaining an adequate replacement part is sometimes the most difficult part of a repair. Fortunately, if you built a project, you should already know where you were able to obtain parts. If you are repairing a unit, or building a project and cannot find exact replacements, this section will provide hints as to adequate substitutions. All major electronic components will be covered separately.

Any resistor can be replaced with a higher tolerance component of the same value. You may also substitute half-watt resistors for quarter-watt ones (or one-watt for half-watt). The major constraint is in the larger physical size of the component. It will work electrically, if it will fit physically.

Capacitors of a higher voltage rating and same capacitance are usually acceptable. With electrolytics, however, don't go too far above the desired rating if you can help it. It is also usually acceptable to use a larger value capacitor in power supply filters, and bypass capacitors. Timing circuits dependent upon the RC time constant function require that changing the capacitor will also require changing the resistor value. Just calculate the RC constant of the needed components. Using the nearest value of capacitance you can find, calculate for the value of resistance that gives you the same time constant. Use the nearest standard value of resistor. Don't try to go too far from the design value; as in some timing circuits, the timing resistor is also a bias resistor. You could create other problems.

Coils and transformers can be substituted in some cases, but care must be used. The power transformer can be substituted with a slightly higher voltage secondary, as was pointed out in Chapter 7, if the supply is regulated. Current ratings of the replacement device can be higher than the original if the larger physical size is acceptable. Band-pass transformers and other specialized components can be substituted occasionally; however, it sometimes takes more skill to determine when a replacement will work than an experimenter might have. If they are junk box parts, or are inexpensive, you could give them a try, if they will fit physically. The best advice here is to proceed with caution, and use identical replacements whenever possible.

Diodes are easily substituted. Clear glass detector diodes are almost always germanium, while power supply diodes are almost exclusively silicon. The parameters you must be cautious of are PIV and forward current rating for power diodes. Choose a device that has a forward current rating that exceeds the demands of the power supply by at least 100%, and a PIV of at least twice the reverse voltage in the circuit. Bias diodes in amplifiers are sometimes critical, and you may have to experiment to find an adequate device. See Chapter 7 for more details on silicon power diodes.

Germanium detector diodes are noncritical in most cases. A 1N34 or 1N60 will work most of the time.

Transistors are easily substituted, given the proper information. Assuming you have tried the cross-reference and cannot find a replacement listed, or the transistor is unmarked, you can try to determine a replacement.

There are six parameters that must be identified when replacing transistors. Each will be discussed, in the order in which they are usually listed in most replacement guides.

1. Identify the material (silicon or germanium) and polarity (NPN or PNP). If you have a single good junction, use the ohmmeter transistor test to determine the polarity of the device. Then, using a 1.5 volt battery, the transistor under test, and a 10-Kohm resistor in series, measure the voltage drop across the junction. Be sure the junction is forward biased (N material to the negative battery terminal). A silicon device will drop 0.6 volts, while a germanium transistor will drop 0.2 volts.

If both junctions are defective, you have a bigger problem. You must test the polarity of bias voltages in the circuit. An NPN transistor will have a positive voltage at the collector, when compared to the emitter. A PNP device will have opposite polarity. Testing for silicon or germanium is more difficult. You may have to test for this experimentally. Most recent equipment uses silicon devices. Check the transistor bias voltage in nearby circuits. Chances are if those devices are germanium, the replacement transistor will be also. Also keep in mind that a germanium circuit usually is more complex and has more bias components than a silicon circuit.

2. Choose a transistor that has a frequency response equal to or greater than the expected response of the circuit. For example, an AM IF amplifier must have a response in excess of 455 KHz. An audio amplifier should have a response in excess of 20 KHz. Do not choose a device that has more than twice the necessary response. The high gain these devices would have at the operating frequency might cause the device to oscillate.

3. Measure the supply voltage and use that voltage as the minimum requirement for collector-to-emitter and collector-to-base voltages. Choose a transistor whose ratings exceed those amounts.

4. Once you have determined the requirements for collector-to-emitter voltage, you can determine the required collector current and power dissipation. The value of load resistor in the circuit will be a good indicator of collector current. Select a value of current that is larger than the current would be if the transistor were saturated, and the resistor would have maximum voltage across it. For example, a 1 Kilohm load resistor and a 6-volt power supply would call for a minimum current of 6 ma. Leave plenty of latitude for this amount. In a similar manner, determine the power requirements. Power is equal to collector current times supply voltage.

5. The transistor gain, or beta, should be looked at next. Typical gain for most single-stage amplifiers is around 100. Amplifier driver stages usually have lower gain than this. Output stages usually have gains lower than 50. High gain preamplifiers can have gains in excess of 400.

6. The last important parameter is packaging. Normally there is nothing critical about case style unless there is a lot of mechanical work to change style. Small signal transistors have different lead configurations, so you must be sure that the replacement transistor is installed correctly. Sometimes this means that you must insulate the leads and rearrange them so that they fit into the proper holes in the circuit board.

Integrated circuits have very few substitutes. If you cannot find an adequate substitute listed in the cross-reference guide, you will have to order an exact replacement from the manufacturer of the unit. Fortunately, most ICs are available from the usual electronics parts suppliers.

Most other components require exact replacements. If parts are unavailable from the manufacturer, repairs are often impossible. Don't forget that with older equipment you might check with electronics repair shops. They might have a junked chassis just like the one you are repairing. The part you need may not be defective, and you will have found a replacement.

The next section contains flow charts that should help you locate a problem in a given unit. If you enjoy the process of troubleshooting, these charts will be a continuing source of information for you.

Troubleshooting Flow Charts

The last section in this chapter contains five valuable flow charts that condense the troubleshooting information you have learned. Use these charts by starting at the top and working toward the bottom. Audio amplifiers and AM receivers are covered. You can use these charts for any electronic equipment. For example, if you have a defective tape recorder, use the chart for an audio amplifier.

About the only types of equipment you cannot troubleshoot are the RF and IF stages in FM receivers. This is because the detector in the signal tracer is AM, and cannot receive the FM transmissions. You are limited to testing audio stages in FM equipment. You can measure the bias voltages of FM IF and convertor circuits, however be cautious. Moving those small wire coils around will change the alignment of the FM receiver, and could render it inoperative. You do not have the proper equipment to restore proper alignment. Unless you are sure you can make a voltage measurement without accidentally moving components, don't.

Again, these charts are most usable for troubleshooting projects that you have constructed, rather than repairing commercial equipment. The procedures are the same, and differ only in degree of complexity. After you gain some experience on your own projects, go ahead and see if you can troubleshoot a commercial unit. The majority of repairs are within the reach of the practiced experimenter.

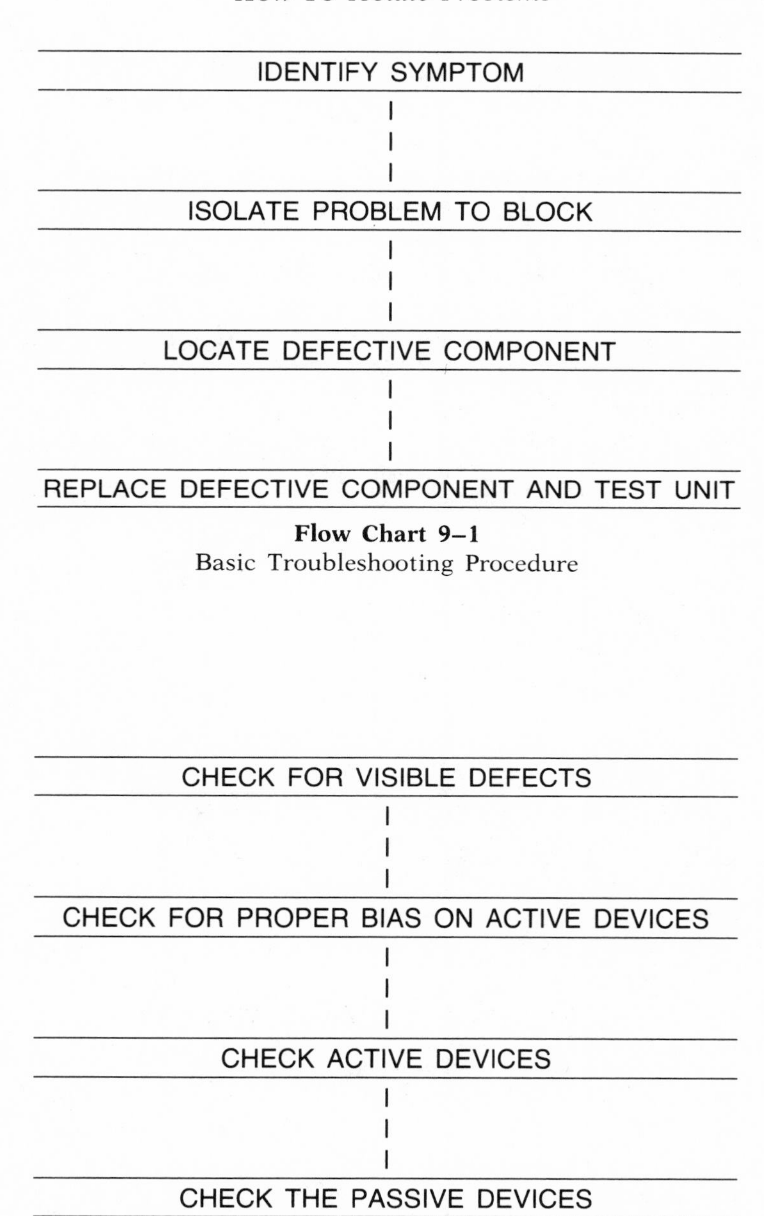

Flow Chart 9–1
Basic Troubleshooting Procedure

Flow Chart 9–2
Locate Defective Components

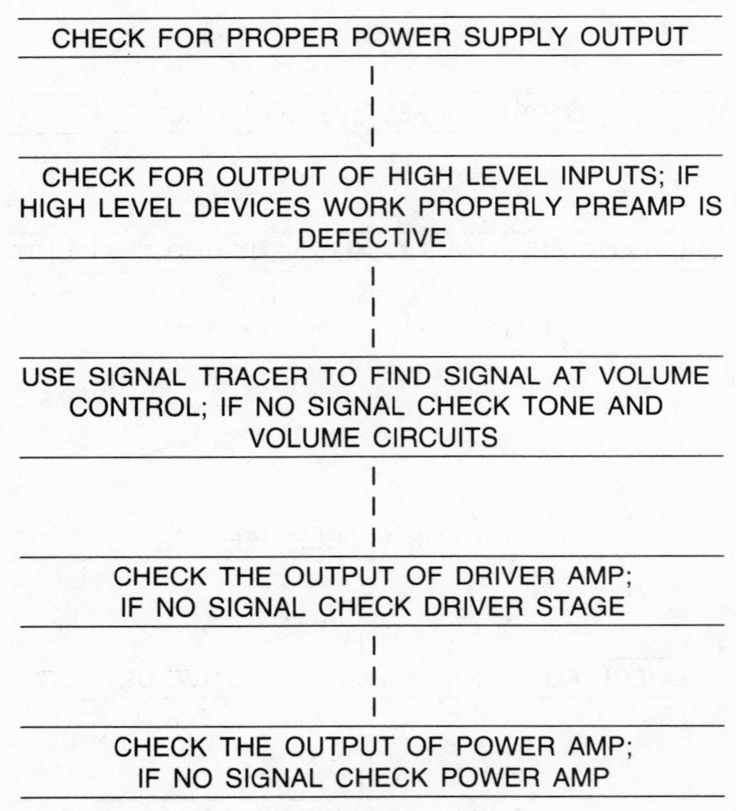

Flow Chart 9–3
Locate Defective Block in Mono Amplifier

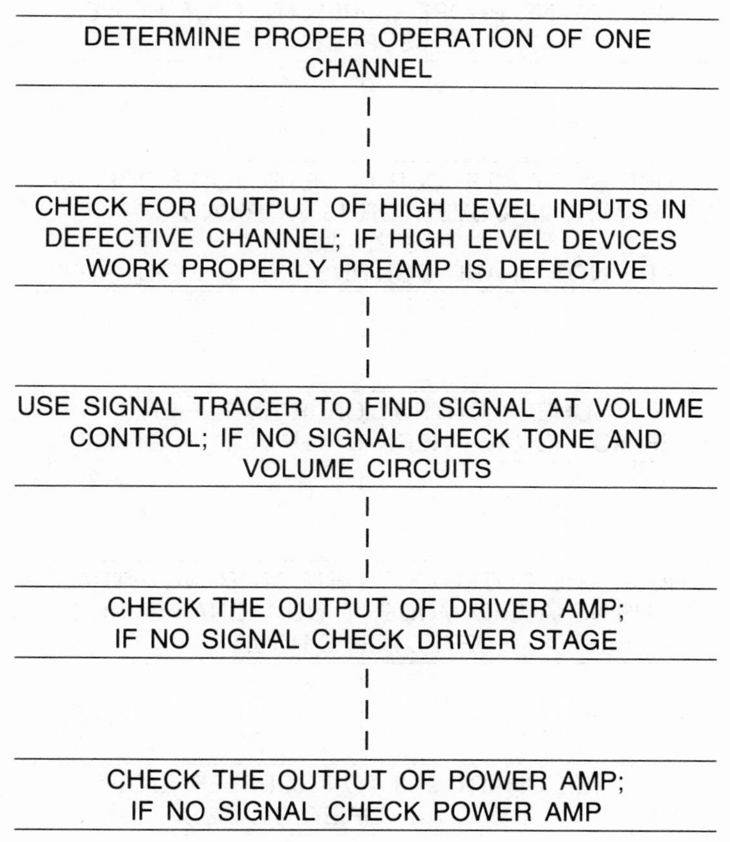

Flow Chart 9–4
Locate Defective Block in Stereo Amplifier

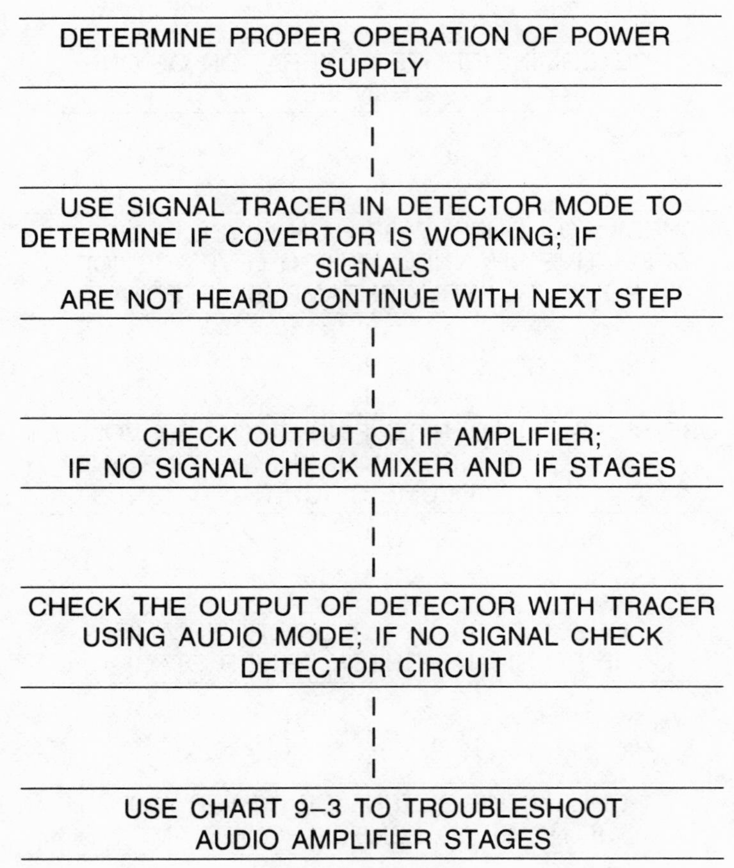

Flow Chart 9–5
Locate Defective Block in AM Radio

Projects That Will Save Time And Money

10

Now it is time for you to put some of your new-found skills to practical use. This chapter contains the instructions and schematics to complete a dozen electronic circuits. They are functional and useful projects around the workbench. Some of the projects are the first stages of more complex projects found later. In other words, the IC audio amplifier is the major circuit in the signal tracer. The regulator circuit can be used in conjunction with the DC power supply to build a regulated power supply.

There are twelve sections to this chapter, each devoted to a single project. Each section is broken down into the following format:

A. Project functional description

B. Circuit description

C. Construction notes

D. Parts substitution guide

E. Troubleshooting tips

By following each of these steps, you will be able to complete these projects with successful results. All projects have been built and tested, and are simple enough so that beginners will have little trouble in their construction.

The following tools are required for maximum ease of construction:

A. Long-nose pliers
B. Diagonal cutters
C. 25-watt soldering pencil (or wire-wrap tool)
D. Drill and assorted bits (if you plan on mounting these projects in a cabinet)

Most of the projects need the following parts in addition to the components listed in the individual parts lists:

A. Perf board (0.100 × 0.100 hole spacing)
B. Solder (rosin core electronic)
C. 20-gauge or 22-gauge wire for component interconnection (or 30-gauge wire-wrap wire) as required
D. Suitable enclosure (if desired)
E. Misc. hardware if mounting in enclosure

The first step in beginning projects such as these is to read through the descriptions of the project operation and construction. Become familiar enough with the projects so that you can easily begin construction and do most of the building by referring only to the schematic. Then gather together all of the components. It is possible to begin building while still waiting for a component that is on order; however, be sure you know that component's actual dimensions. If you do not, you may not leave enough room for it, which will cause you headaches later. All of these projects could have been built on a PC board, however in the interest of ease of construction, perf board is used. Double- and triple-check each connection for proper placement and good solid electrical connection. Most electronic projects fail, not because of incorrect wiring, but because of poor soldering, loose connections, etc. A professional attitude toward this portion of the hobby will go a long way in keeping errors to a minimum.

If you have a prototype development board, don't hesitate to build the projects on there first. That will allow you to troubleshoot any problems you may encounter before actually committing components to solder. It is not actually required, but there is an advantage to building the power supplies first. They can provide the power required to operate the later projects. Battery power is also an option in most of these circuits, and you may find it convenient for portable operation.

If you run into trouble with a project, cannot seem to find any wiring errors, have used the troubleshooting tips, and it still doesn't work, don't give up yet. First, set the circuit aside, and look at it again a few hours later, or preferably the next day. A fresh look at the project might let you see something you have constantly overlooked. If you still cannot find the problem, give the circuit to a friend and let him or her look it over. It is possible to continually overlook a simple missed or incorrect connection, even after checking it over several times.

Gather everything together, settle down at your work station, and let's begin building.

RC Substitution Box (Figure 10–1)

Project Description:

The RC (resistor-capacitor) substitution box is a useful device that provides three functions around the experimenter's bench.

A. It can replace suspected defective components with good ones.

B. It can allow the experimenter to change component values to test the effects on a given circuit.

C. It can supply a temporary replacement part in a circuit until a permanent component becomes available.

The project is a good beginner's project because it is relatively simple to construct and requires no power supply. It also will give plenty of practice in soldering noncritical components. After completing this project successfully, you should be able to complete the other projects with few problems.

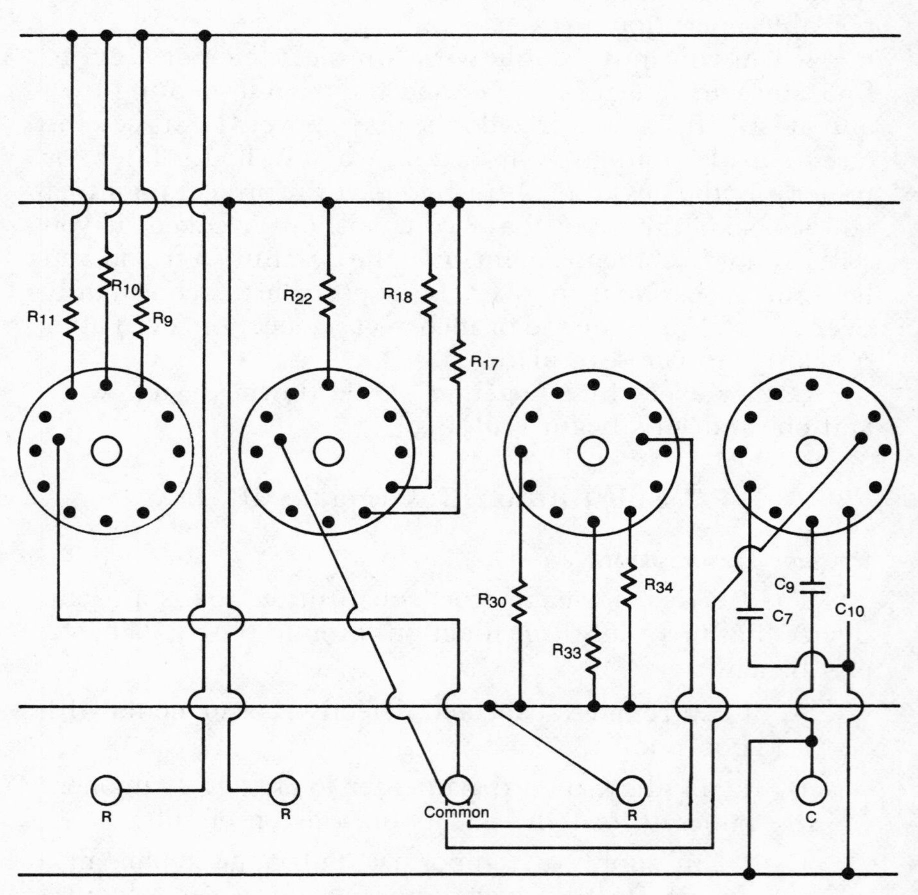

Figure 10–1
RC Substitution Box

The box contains 36 resistors and 12 capacitors, but if you wish you could increase the number of capacitors by eliminating a resistor bank, or you could add extra banks for more capacitors or resistors as desired.

The operation of the device is simple. Insert one test cable into the common jack, and the other cable into the correct component jack. Select the proper switch position, then insert the test leads into the circuit being substituted. The component inside the box will substitute for the faulty or missing component. This useful accessory will save you hours of soldering time while trying out substitute components.

Circuit Description:

Refer to the schematic diagram while reading the following description. There are four banks of single-pole twelve-throw switches. All of the poles are connected to a common terminal, and each throw is connected by a bus to an individual jack. If you select an individual switch position, only one component at a time is connected to the common and switch output jacks. Connecting leads to these points allows the experimenter to replace the internal components with those from an external circuit.

Construction Notes:

In order to keep the schematic easily readable, only three components are shown on each switch. The switches are wired clockwise from the back of the switch, starting with the largest resistor or capacitor value. This provides a clockwise reading from low to high value when using the box.

There are only a couple of cautions involved in this circuit. Be sure to connect the electrolytic capacitors with their positive leads to the capacitor bus. An incorrect connection could cause the capacitor to overheat and explode. Do not use the capacitors in high-voltage circuits as the capacitor voltage ratings are not high enough.

The project is best constructed in a plastic project box such as Radio Shack's 270-224. You can mount the circuit on the cover and install it in the box when you are done. If you choose a metal box, or one with a metal cover, be sure the bus wires are insulated from the case. The bus wires are supported at each end by ordinary terminal strips.

The resistors can be 10% half-watt values, or for greater precision at little extra cost, you can use 2% flameproof resistors. You may have to order those through a radio-TV repair shop, as many consumer electronics places do not stock flameproof types. One-watt resistors are also usable, and will allow you a little more versatility when substituting values in power circuits. It would probably be a good idea to have one-watt resistors for 100 ohms and smaller values.

Capacitors above 1 microfarad are electrolytic, and as was stated earlier, must be connected with their positive leads toward the bus wire. Capacitors with at least a 50-volt rating should be used, and 100-volt devices would be better if you can find them. Values under 1 uf. can be mylar or disk or intermixed. If you plan on using the box to find component values for timing circuits, use temperature-stable capacitors.

Parts Substitution Guide:

Switches will probably be the hardest components to locate. You can substitute with any single-pole multithrow switch you desire. You will have to change only the total number of switches desired. Commercial substitution boxes use multipole switches ganged to a single shaft. If you have some old equipment around, you may find a suitable switch or two.

The binding posts may be eliminated if you want to run test lead wire to the common terminal, and a lead to each bus wire. It is more convenient to have a jack and plug assembly, especially if you choose a design that matches existing equipment. You will be able to interchange leads with other equipment as desired.

All capacitors should be 50 volts DC minimum.

Misc. Parts: Four single-pole twelve-throw rotary switches (Radio Shack #275-1385), five binding posts and hardware, cabinet, wire, knobs, mounting hardware, etc.

Parts List

Resistors

Resistor	Ohms
R1	10
R2	15
R3	27
R4	33
R5	47
R6	68
R7	100
R8	150
R9	180
R10	270
R11	330
R12	390
R13	470
R14	560
R15	680
R16	1000
R17	1500
R18	2200
R19	3300
R20	3900
R21	4700
R22	5600
R23	6800
R24	10K
R25	15K
R26	27K
R27	33K
R28	39K
R29	47K
R30	56K
R31	100K
R32	150K
R33	270K
R34	470K
R35	1M
R36	2.2M

Capacitors

Capacitor	Mfd.
C1	.001
C2	.005
C3	.01
C4	.05
C5	.1
C6	.22
C7	.5
C8	10
C9	50
C10	100
C11	500
C12	1000

Troubleshooting Tips:

To test your circuit, connect an ohmmeter to the common terminal and each bus wire in turn. As you rotate the switch, you should be able to read the correct resistance at each position. An open indicates an improper solder connection to that resistor; a wrong value indicates a resistor out of place or a short between two posts.

To test the capacitor bank, put the ohmmeter on the highest scale. As you rotate from the smallest to largest capacitor, the charge current should get increasingly higher. If you cannot remember how to check a capacitor with an ohmmeter, see Chapter 2.

DC Power Supply (Figure 10–2)

Project Description:

The basic power supply was covered in Chapter 7. The project applies the rules to eventually become a 5-volt regulated supply. The unregulated supply puts out between 7 and 10 volts, depending upon the transformer chosen and current demand. The next project adds the 5-volt regulator circuit, which will allow the supply to power projects that require a 5-volt regulated source.

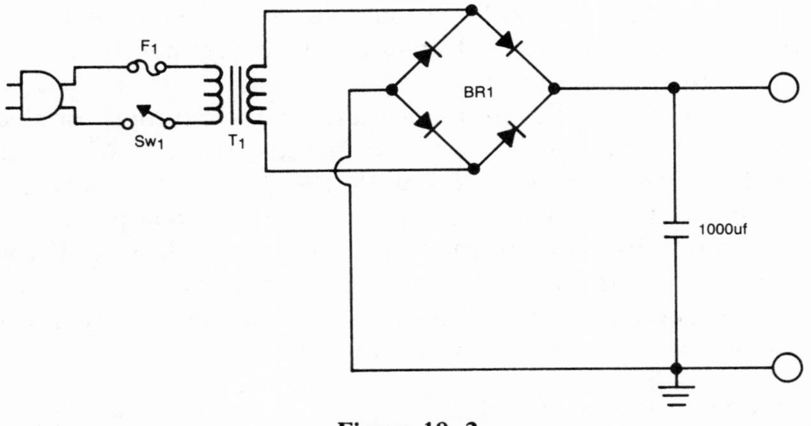

Figure 10–2
Power Supply

Circuit Description:

The supply circuit uses a full wave bridge with a capacitor input filter. Design values came from Chapter 7. See that chapter for a more detailed description of circuit operation. It will deliver 7 to 10 volts at 1 amp.

Construction Notes:

The bridge rectifiers and filter capacitor can be mounted on perf board, but choose a large enough board to contain the regulator components in the next project, if you plan to expand it. Mount the transformer in a project cabinet, or other suitable enclosure. The perf board may be mounted inside the cabinet by using spacers on long bolts. The spacers should be long enough to keep the board from making contact with the cabinet. If you are mounting it in a metal cabinet, use a three-wire AC cord. Connect the green wire in the cord to the metal chassis. If you are never going to use the supply to deliver a negative voltage, or if you are going to use it with the regulator circuit, connect the minus binding post to ground, via the chassis or three-wire cord. Be sure you use the supply with a three-wire grounded outlet. Improper grounding can destroy the sensitive ICs being used in the later projects.

If you are going to have long test leads to connect your project to the binding post, install a 0.1 ufd. capacitor across the output of the supply.

If you want to mount a voltmeter in the cabinet, connect it to the binding posts from the inside. An ammeter could also be connected in series with the positive or negative binding post. Another alternative is to install test posts that allow you to insert a voltmeter or ammeter as desired. If you choose to install posts for reading current, remember that you will need a shorting bar or wire to connect between the posts when the meter is not connected.

Parts Substitution Guide:

Parts not included on the parts list are identified on the schematic. Two bridges are specified, one a bridge assembly, the other discrete diodes.

Part	Quantity	Name	Radio Shack	Sylvania
F1	1	0.25 amp	270-1270	NA
T1	1	6.3 V 1.2 A	273-050	NA
Br1	1	1.4 A 50 PIV	276-1151	ECG 5304
Br1	4	3 A 50 PIV	276-1146	ECG 125
C1	1	1000 ufd 16 V	272-958	NA

Misc. parts: Hardware, connectors, binding posts, AC line cord and plug, fuse holder.

Troubleshooting Tips:

After completing the project, connect an ohmmeter between ground and both positive and negative points. There should be infinite resistance. Next measure the resistance between the positive and negative posts. On the high-resistance ohmmeter scale, you should see an increasing resistance until the filter capacitor charges completely. If the resistance is not infinite, after the capacitor charges, you have a problem and should recheck your wiring.

The bridge assembly is easier to install than the four diodes, but if you have used the four diodes, they are easily checked for proper installation. The positive lead from the filter capacitor must connect to two banded diode ends, or the + on the bridge assembly. The capacitor negative lead must connect to two unbanded diode leads, or the minus on a bridge assembly. The banded/unbanded junctions that are left both go to the transformer secondary. This point on most bridge assemblies is marked by a tiny sine wave. Polarity of the transformer leads doesn't matter.

If the fuse blows with no load connected to the supply, recheck the transformer and bridge rectifier wiring for solder bridges, or incorrectly installed components. Low or no output voltage indicates an open circuit, unless accompanied by heating of the components. If the components run warm with no load connected, be sure to check the polarity of the filter capacitor, and recheck for solder bridges, or other shorted components.

If you have no output voltage, measure to see if you have AC at the secondary of the transformer. If no AC voltage is found, check the primary winding, fuse, and AC line wiring. If it is present, check the bridge and filter wiring, and the wiring to the binding posts.

Power Supply Regulator (Figure 10–3)

Project Description:

The regulator circuit shown here interfaces with the power supply in the last section. The output of this regulator is +5 volts DC. The 5-volt source is commonly used for

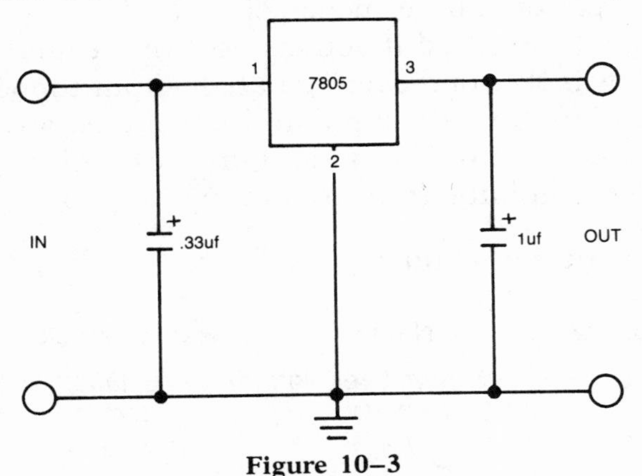

Figure 10–3
Regulator Circuit

digital integrated circuit projects and other general purpose work. You will need a 5-volt supply to construct the LED digital counter circuit. You may desire to put an extra set of binding posts on the output of the supply—one set connected before the regulator to provide unregulated voltage, and one set after the regulator, to supply the regulated 5 volts.

Circuit Description:

The circuit uses a 7805 voltage regulator IC. The simple device can deliver up to 1 amp at 5 volts. It uses integrated circuit technology to provide internal shutdown should the device overheat. It will accept an input voltage of up to 35 volts.

Construction Notes:

The project will be easy to construct, if you left enough room on the perf board in the power supply. Alternately, you could mount the regulator circuit on its own board, and enclose it in the same cabinet.

The regulator IC must be heat sinked, if it is to work to maximum efficiency. If you used a metal cabinet, you may ground the heat sink tab directly to the cabinet. The surface area of the cabinet should act as an adequate heat sink. Another alternative is to connect it to a small metal heat sink, either homemade, or commercial. It should have at least a couple of square inches of area.

C1 may be omitted if you connect the regulator circuit within a few inches of the unregulated output connections. It is required any time the regulator is mounted any distance from the main supply. The major purpose of this capacitor is to keep the regulator from oscillating.

Parts Substitution Guide:

Part	Quantity	Name	Radio Shack	Sylvania
IC1	1	7805 5 volt regulator	276-1770	ECG-960
C1	1	0.33 uf 50 V	NA	NA
C2	1	1 uf 16 V	272-1419	NA

Misc. parts: Hardware, heat sink, etc.

Troubleshooting Tips:

If this circuit doesn't work, check the wiring, and be sure you have the IC mounted with the right side up. The center pin of the IC is ground, and does not have to be connected if the heat sink tab is grounded. Low or no output voltage with no load connected as before indicates there is an open. If the IC is getting warm to the touch, it could be shutting down. When that happens, you may have a shorted connection on the output line to ground. An unloaded IC will not run very warm. If all wiring is correct, measure the input voltage to the IC (pin 1). If it is there, measure the output (pin 3). If you do not have 5 volts there, the integrated circuit is defective.

Dual Voltage IC Supply (Figure 10–4)

Project Description:

The dual voltage supply is used to power many integrated circuits. The supply will deliver +12 volts and −12 volts at 1 amp. Again, the design is based on parameters from Chapter 7.

None of the projects in this book requires the negative side of the supply. You could choose to build only the positive going supply if you wish. Just use a single 12 V transformer, without a center tap, and omit the negative supply parts. If you plan on further experimentation, you may find the dual supply extremely valuable.

Circuit Description:

The dual supply, though it uses a bridge rectifier assembly, and looks similar to a bridge, operates more like a conventional full wave rectifier. The center tap on the transformer converts the bridge to two diode pairs, each operating as a conventional full wave circuit does. One pair delivers positive voltage out, while the other pair delivers the negative voltage.

Construction Notes:

Finding the transformer may be the most difficult part. It is a 24-volt, 2-amp transformer with center tap. If you

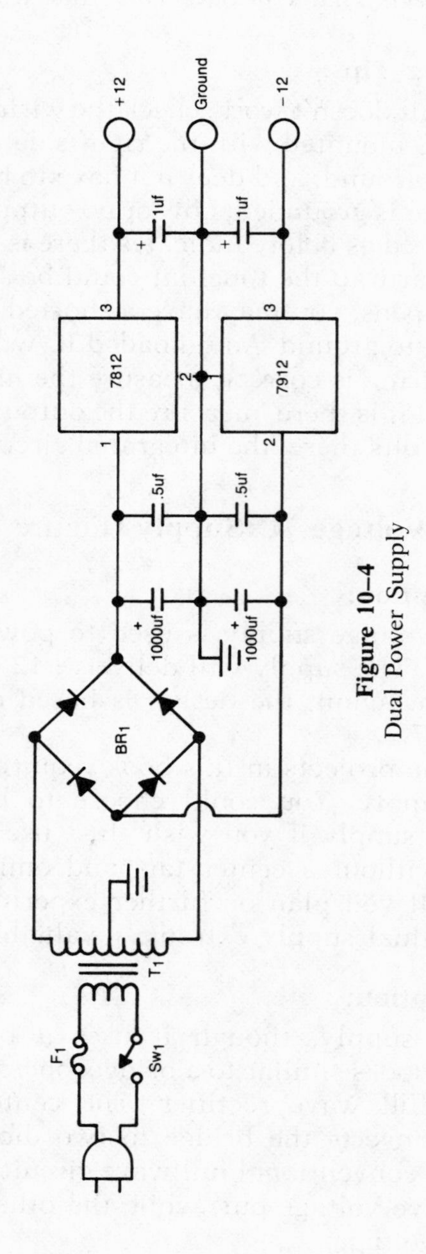

Figure 10–4
Dual Power Supply

elect to build only a positive going supply, select a 12-volt, 1-amp or better transformer without a center tap. As in the earlier supply, you may use a bridge rectifier or discrete diodes. The filter capacitors must be mounted with respect to proper polarity as they are also reversed in the negative going supply.

One caution is in order regarding the negative voltage regulator: The heat sink on the device is not ground, as it is on the positive regulator. You must isolate the tab on that IC from ground and from the tab on the other IC. You can mount it to an isolated piece of metal, or use a mica insulating washer and nylon screw to bolt it to a grounded heat sink if needed.

Parts Substitution Guide:

Any parts not listed in the list below are identified by component value on the schematic. All capacitors are 16 volts or greater. All values above 1 uf. are electrolytic.

Part	Quantity	Name	Radio Shack	Sylvania
IC1	1	7812 12 volt regulator	276-1771	ECG 966
IC2	1	7912 – 12 volt regulator	NA	ECG 967
Br1	1.	1.4 amp 50 PIV	276-1151	ECG 5304
T1	1	25.5 v 2 amp transformer	273-1512	NA

Misc. parts: Perf board, fuseholder, enclosure AC line cord assembly, hardware.

Troubleshooting Tips:

As with the previous projects, check the wiring thoroughly. Refer to the two previous troubleshooting tip sections for checking out the regulated and unregulated parts separately. To assist in determining whether the problem is in the supply or regulator, disconnect the input of the regulator, and check the output of the supply. If it is all right there, check the regulator circuit. Remember to double-check the polarity of the diodes and capacitors, especially in the negative going supply.

Relay or LED Driver (Figure 10–5)

Project Description:

The relay driver is designed around the circuit values described in Chapter 8. Review how to design a transistor switch for more detail on the process.

The relay driver will be used any time you want to remotely switch a device. The device is connected to a relay, and control of the relay is accomplished by connecting a positive voltage to the base of the transistor. The driver can be powered by battery or one of the supplies built earlier.

The LED driver is calculated using a 9-volt transistor battery, while the relay driver uses a 12-volt source. Chapter 8 explains how to change the parameters for other supply voltages.

Circuit Description:

The circuit uses a bias voltage on the base of the transistor to control current flow in the collector circuit. The

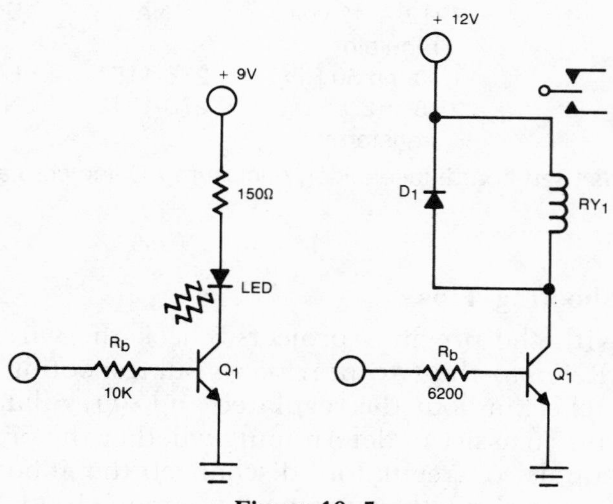

Figure 10–5
Relay or LED Driver

base current is dependent upon Rb, which is chosen to saturate the transistor. When bias is applied to the base, collector current flows and switches the LED or relay on.

Construction Notes:

The circuit can be wired on perf board, or any other convenient way. The value of Rb is calculated to be driven from a 9-volt supply. The input of Rb can be connected to +9 volts through a switch, or through the output of the universal timer you will build later.

Parts Substitution Guide:

Parts not in the list are identified on the schematic.

Part	Quantity	Name	Radio Shack	Sylvania
Q1	1	2N2222 transistor	276-2009	ECG 123A
LED1	1	light emitting diode	276-041	ECG 3007
RY1	1	12 volt 160 ohm coil 10 amp 125 v contacts	275-218	NA
D1	1	1N4001 Silicon diode	276-1101	ECG 125

Troubleshooting Tips:

The first step if this circuit doesn't work is to recheck the wiring. Be sure that the transistor is connected correctly. It is easy to get the three leads mixed up.

If the transistor stays on, measure the input voltage. The input side of Rb should change from +9 volts to 0. If it is changing, measure the collector of Q1. That point should be the opposite state of the input voltage. If it is not switching, it will be either 9 volts or 0 volts all the time. If it is 0 volts, there is an open between the collector lead and the positive power supply, or there is a short to ground, either through the transistor (which would be defective) or externally. If the voltage is positive all of the time, the transistor is open, or not grounded properly. Also check the wiring to Rb, if that is the case.

If the transistor stays on, check to see that the input voltage to the Rb is changing state. You will find the collector voltage at or near 0 volts in this case. Using a clip lead, connect the base and emitter leads together. The transistor should immediately shut off. If it doesn't, either it is defective, or there is a wiring error.

The third possible condition is the transistor switches from 9 volts down to a voltage higher than 0. As you will recall, this means the transistor is not saturated. Your transistor probably doesn't have quite enough gain for the size of base bias resistor. Lower the value of Rb until the collector voltage drops solidly to 0. Here is a good application for that RC substitution box you built.

Universal Alarm or Noisemaker (Figure 10–6)

Project Description:

This device is an extension of the previous project. By substituting a siren, buzzer, or other noisemaker, you can make a suitable alarm. Switch Sw1 can be an ordinary toggle switch for testing, and it can be changed to a mercury switch for a position alarm. If the switch is tilted, the alarm will sound. A magnetic switch can be used as an alarm. When by a closed door, for example, the switch could remain open. When the door opens, the noisemaker will sound. A

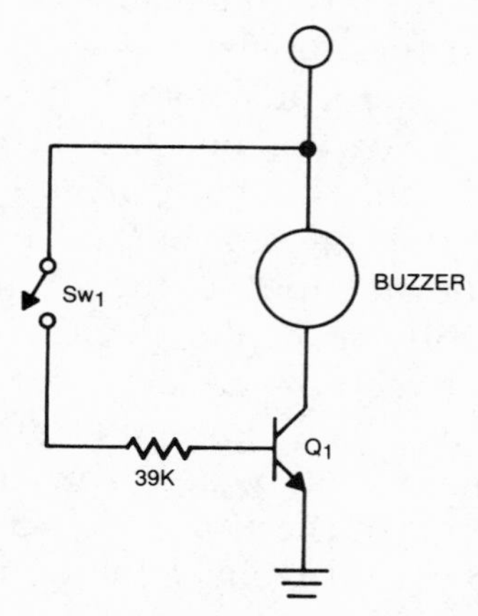

Figure 10–6
Alarm or Noisemaker

moisture detector could be made by positioning two bare wires very close together. If water rises high enough to cause current to flow between the two wires, the alarm will sound. You will have to experiment with the length of wiring and spacing to make that application work.

Circuit Description:

Like the last circuit, this is just another application of a transistor switch.

Construction Notes:

There is nothing at all critical about the circuit components in this transistor switch. Use any kind of wiring you desire.

Parts Substitution Guide:

Q1 2N2222 transistor Radio Shack 276-2009 Sylvania ECG 123A

Noisemaker Radio Shack 273-060 Piezo electric buzzer

Sw1 Any switch desired for application

Misc. parts: Battery, holder, mounting hardware, perf board, enclosure, etc.

The noisemaker can be changed, using any device requiring 9 volts. To recalculate, double the device current and divide by 100. Divide the result into 9. The result will be the amount in ohms of Rb.

Troubleshooting Tips:

This circuit operates in exactly the same way as the relay driver discussed earlier. Refer to that section for a detailed description.

Stereo Audio Preamp (Figure 10–7)

Project Description:

The transistor is about to be replaced by the integrated circuit. In this and the following projects, you will be introduced to some new devices that are cheaper and easier to work with than equivalent transistor circuits.

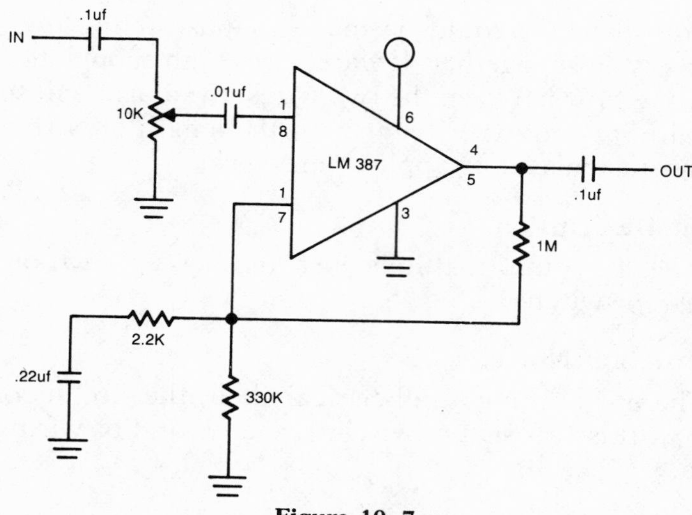

Figure 10–7
Stereo Audio Preamp

The project shown here is a simple stereo/audio microphone preamplifier. It can be used to connect any low impedance microphone to a power amplifier. The IC contains two identical amplifier circuits, and if you desire, you can connect them both for stereo operation. One application of this device would be a line driver—preamplifier mounted close to the microphone, which will amplify the signal before it starts a long conductor run to the PA system.

To experiment with this circuit, you can connect the output of this amplifier to the tape or high-level phono input of your stereo. You can also connect it to the input of the IC audio amp you will build as the next project. That amp is very high gain, however, so keep the volume low, or you may damage the amplifier's speaker. For an input device, you can use any inexpensive cassette recorder microphone. In a pinch, a speaker will also work, or even a headphone speaker will make a tolerable microphone.

Circuit Description:

The circuit is basically an LM387 dual preamp. It is a high-gain amplifier that is easy to bias, and provides easy interfacing to external circuits.

Audio input is at the left through the 0.1 ufd capacitor. The variable control is an input level adjust. This component could be put at the output if desired. Input to the IC is via the 0.01 uf. capacitor on pins 1 and 8. (Pins 1,2, and 4 are one channel; pins 5, 7, and 8 are the other.) The bias components connected to pins 2 and 7 set the gain of the device. Amplified audio is taken from the output pins 4 and 5, and coupled out of the circuit via the 0.1 uf. capacitor on the right.

Power is obtained from a 12-volt source. Though I have never tried it, it would also probably work acceptably with a 9-volt battery.

Construction Notes:

The LM387 is a high-gain amplifier. As such, it needs to be handled carefully so that feedback does not occur. Wire the circuit on perf board, keeping inputs as far as possible away from outputs. Any sign of instability or oscillation means you should try to reposition components. The device is relatively stable, and should give little trouble, especially if leads are kept reasonably short.

Parts Substitution Guide:

Part	Quantity	Name	Radio Shack	Sylvania
IC1	1	LM387 IC	NA	ECG 824

Misc. parts: Enclosure, perf board, input and output jacks, IC socket, etc.

All other components will tolerate some deviation from stated value. Don't change too many components, however, because you could run into problems.

Troubleshooting Tips:

Check and double-check your wiring. It is easy to get confused when wiring ICs. From the bottom, ICs are numbered in a clockwise direction from the upper left, around to the lower right. Measure the power supply voltage. If the voltage is present, use a signal tracer to see if signal is getting to pin 1 (or 8). A signal there, with no signal at the output, points to either a defective IC or problems in the bias network. Before condemning the IC, check the bias network at pin 2 (or 7). No problems here indicate a defective IC.

IC Audio Amp (Figure 10–8)

Project Description:

The audio amplifier is a versatile item to have around the house and shop. The high-gain amplifier that is described in this section is popular in auto radios and CBs. No preamp is required to drive a signal from low level, such as a phonograph or tuner to 5.8 watts output. The amplifier, in addition to being easily applied to other projects, is the heart of the signal tracer. In fact, there are only a few components required to make this circuit work properly as a tracer.

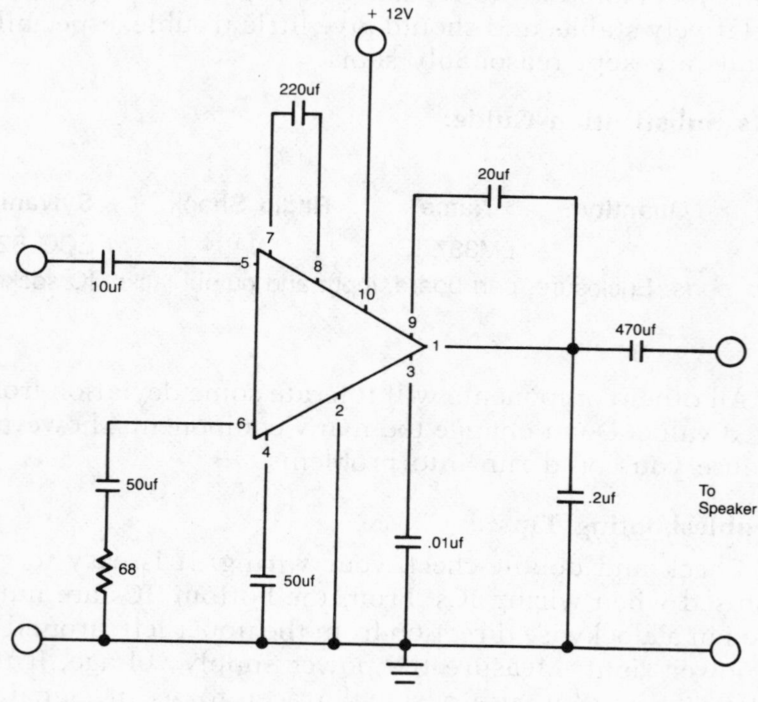

Figure 10–8
IC Audio Amplifier

Circuit Description:

The circuit is designed around a Japanese high power integrated circuit amplifier. The BA-521 is designed to act as an audio power amplifier. Audio is applied to the gain control, where it passes to pin 5, the IC input. Gain of the device is set by the single resistor, R1. The value of the resistor can be increased to about 330 ohms. A tremendous change in gain will be noted as that resistor is changed. Amplified audio is delivered to the speaker through a 470 uf. capcitor, via pin 1. Speaker impedance can be 4 or 8 ohms.

The device is at home in automotive applications, and is designed to run with a power voltage between 12 and 14 volts. No power supply is needed if you connect the power connections through a 1-amp fuse to a cigarette lighter plug.

Construction Notes:

Any suitable enclosure is adequate for the BA-521. One option for an enclosure is to buy a small speaker in an enclosure. Install the amplifier inside the speaker enclosure, putting a jack on the speaker back to connect the input component. Two of these, coupled to a multi input stereo preamp would make an inexpensive ministereo system. If you plan on using the circuit for a signal tracer, leave room on the perf board for a handful of other components. Also allow for mounting of a switch, to be explained later.

The high-gain amplifier IC can have feedback and oscillation problems, so be sure to keep the input leads away from the output circuit, especially where they connect to the jack assembly. As with all low-level audio circuits, use shielded cable on the input leads.

In experimenting with this circuit, we found that unregulated supplies tend to make the device unstable. If the circuit oscillates, distorts, or operates weakly, try hooking it up to a car battery, or two six-volt lantern batteries hooked in series. If the problems disappear, the power supply needs more filtering, or better regulation.

The IC needs a heat sink, especially if it is driven hard. Do not operate it without at least three square inches of heat sink, more if you plan on continuous duty use. A commercial heat sink is probably called for here. Be careful not to let the heat sink touch ground, as the tab on the IC is not ground. If you must ground the heat sink, use insulating mica and nylon screws to secure the IC to the heat sink.

Parts Substitution Guide:

Part	Name	Radio Shack	Sylvania
IC1	BA-521 audio	IC276-704	ECG 1165

Misc. parts: Enclosure, 4-ohm to 8-ohm speaker, input and output jacks, perf board, misc. hardware, etc.

Troubleshooting Tips:

Be wary of the output temperature while testing the circuit as too much heat will destroy the device. If the amplifier doesn't work, first check for proper power supply voltage. If that is correct, use a voltmeter to measure pin 1. It should be one-half the supply voltage. If it is not, check for shorts in the output wiring. If there are no wiring errors, and the output voltage is not half the supply voltage, the IC is probably defective.

The only other problems you might run into were described, and solutions were covered in construction notes, above.

Signal Tracer (Figure 10–9)

Project Description:

Chapter 9 demonstrated a technique for repairing audio and radio circuits using a signal tracer. The power amplifier you built in the last chapter is the heart of the tracer. It will work to trace audio circuits, just as it is, by installing a test probe with a shielded cable. The circuit will not troubleshoot RF circuits, however. This project completes the construction of the tracer by adding an RF mode. You will be able to switch the circuit from audio to RF for troubleshooting AM radios.

Another application for this versatile device is as a remote public address amplifier. Install a microphone in the audio input jack, apply power, and you are ready to go.

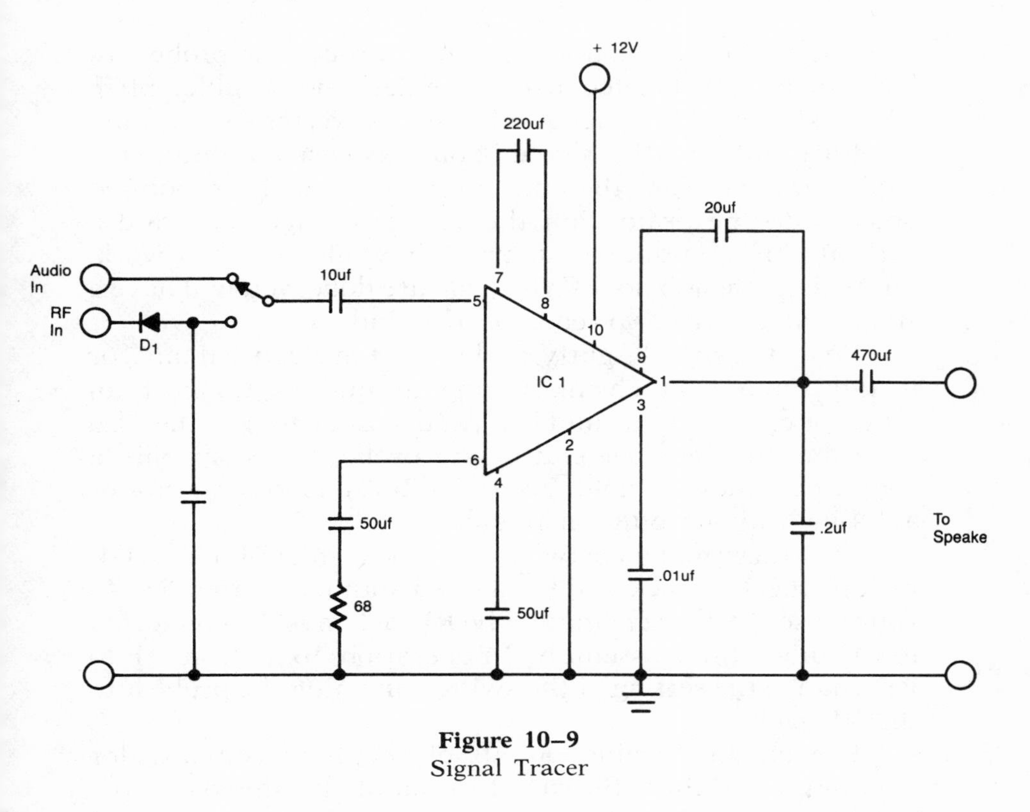

Figure 10–9
Signal Tracer

Circuit Description:

The audio amplifier input in the tracer is connected to a switch, which allows selection of AF or RF modes. The RF mode connects a diode detector and RF filter capacitor to the audio input. When the switch is set in the RF position, signals from this diode are sent to the amplifier.

In the audio mode, the audio probe is connected. You will be able to move through audio circuits; any signals appearing at the test points will be heard in the tracer's speaker.

Construction Notes:

The audio circuit was covered in the last project, so we will look only at the new circuitry.

You will need to build a shielded probe. The probe will have to be constructed from shielded audio cable. Strip about six to eight inches of outer covering away, and carefully unbraid the shield. You can easily unbraid the cable, starting from the end, by inserting a sharp pointed object into the braid. Pull the object through the braid a little at a time, and it will unbraid the cable. It is slow work, but will get the job done. When you are done, you will have a single cable with two leads on one end.

Wrap the shield tightly, and wrap it in electrical tape, or install some heat shrink tubing around it. Connect an alligator clip to the ground lead, and a probe to the other. Do the same to the other end of the probe, but strip only a couple of inches of shield. Install an RCA phono connector or a 1/4-inch phone plug on this end.

The tracer will have two input jacks, either phono or 1/4-inch phone. One jack will be marked audio, the other RF. As you can see by the schematic, the RF jack must be connected to the diode filter assembly. In operation, to go from AF to RF, change the setting of the switch, and plug the probe into the RF jack.

Use shielded cable on all of the input circuits for maximum stability. Be sure to ground the shields to the amplifier ground on the circuit board.

If you want, you can build an AC power supply in the same cabinet as the tracer. Remember the requirement of good regulation in any design you use. As before, a pair of lantern batteries in series will power the device. If you put an external power jack on the back of the tracer cabinet, you could have a cigarette lighter cable for portable operation.

If you plan on an internal supply, or battery operation, you will probably want to install a power switch on the tracer. One way to install a switch is to purchase a variable resistor with a switch mounted on the back. The dual component will act as your gain control and power switch.

Parts Substitution Guide:

For diode D1, any general-purpose germanium diode will work. Type numbers to look for include 1N34 or 1N60. The capacitor can be mylar or disc. These parts could be found in an old transistor AM radio.

Misc. parts: Shielded cable, connectors, probes, power supply components, etc.

Troubleshooting Tips:

If you built and tested the amplifier in the last section, you should have little trouble with the RF portion. No audio from the audio probe indicates that wiring to the probe or AF-RF switch is incorrect. Unplug the probe and use an ohmmeter to test the probe continuity. There should be infinite resistance between the probe tip and ground clip. There should be zero ohms between the probe and center or tip of the plug. There should also be zero ohms between the ground clip and the outer shield or body of the plug.

The RF section can best be checked using an RF signal generator. The next best test is on an AM transistor radio. Using an AM radio, rather than AM-FM, will make it easier to find a good test point. If the radio you are using is not battery operated, remember the cautions about using an isolation transformer. A battery-operated radio is best. Locate the small square metal adjustable IF transformers. These components will have small screwdriver slots or hex-shaped holes in their center.

Turn the radio on to a strong local station, and turn the volume to minimum. Use the probe to check the base and collector leads of any transistors near those transformers. You will be looking for an IF amplifier transistor. When you have found a good signal point, the receiver audio will be heard in the tracer speaker. Once you verify that the tracer works, experiment with the other transistors. Transistors after the volume control will be audio devices, and you can put the tracer in the audio mode. If you turn the volume control down, there will be no audio signals until you turn it up again.

Spend some time practicing and using the tracer. You will find it can be especially useful in audio circuits. The major limitation of the tracer is that it lacks an FM demodulator. You will not be able to use it in the RF portions of FM circuits; however, it still can be used to troubleshoot the audio stages.

IC Audio Generator (Figure 10–10)

Project Description:

The audio generator can be used in addition to a signal tracer. It will generate a variable frequency tone that can be injected into the amplifier under test. By starting from the speaker and working towards the volume control, you will be able to spot dead stages in an audio amplifier.

Circuit Description:

The circuit introduces another popular integrated circuit, the 555 timer. Although it has been around for a few years, there are few devices that can replace it. The circuit is an astable multivibrator. Translation: It is a square wave oscillator. The circuit generates a square wave, which will drive a speaker directly. The only two adjustments needed are gain and frequency. The frequency range of the oscillator using the components specified is from less than 300 hertz to greater than 10 Khz. The large range makes it difficult for a single 500K control to set a precise frequency. If you want to set frequency accurately, use a multiturn potentiometer.

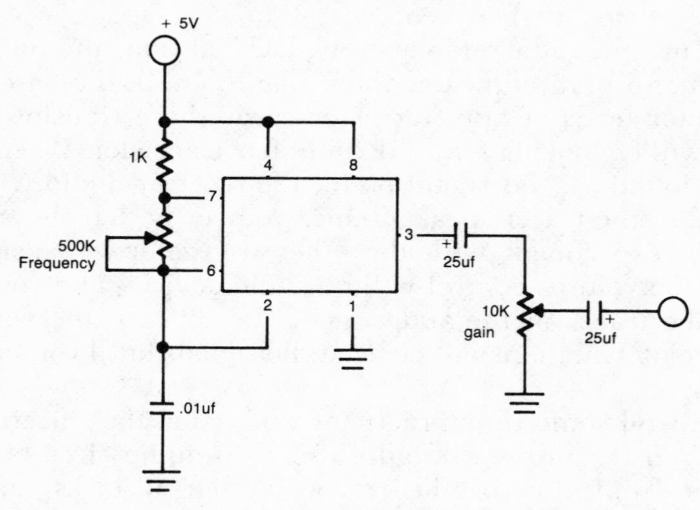

Figure 10–10
Audio Generator

The power supply requirements are not critical—any voltage between 4.5 and 18 is acceptable. A 9-volt battery could be used instead of a 5-volt DC supply.

Construction Notes:

The circuit seems to be totally noncritical. Changing the capacitor at pin 2, or the resistors connected to pins 6 and 7 will change the frequency of oscillation. If you desire a fixed frequency generator, you can eliminate the frequency adjust and substitute a fixed resistor that gives you a pleasing tone.

Any suitable enclosure is adequate, and the output circuit could use the same jack as the signal tracer. This would allow you to use the same probe for the tracer and generator.

Parts Substitution Guide:

Part	Name	Radio Shack	Sylvania
IC1	NE555 timer IC	276-1723	ECG 955

A lower value potentiometer could be used for frequency adjust, but it will limit the range of output frequencies.

Misc. parts: Enclosure, perf board, power connectors, battery, power supply, generator probe, output jack, misc. mounting hardware, etc.

Troubleshooting Tips:

There is little to go wrong with this circuit. If it doesn't work and the power supply is delivering voltage to pins 4 and 8, recheck your wiring. If there are no wiring errors, and the power supply is operating, the IC is probably defective.

LED Digital Counter (Figure 10–11)

Project Description:

The digital counter circuit was included in the project section to introduce you to the world of the digital IC. The circuit demonstrates how easily counter circuits can be fabricated.

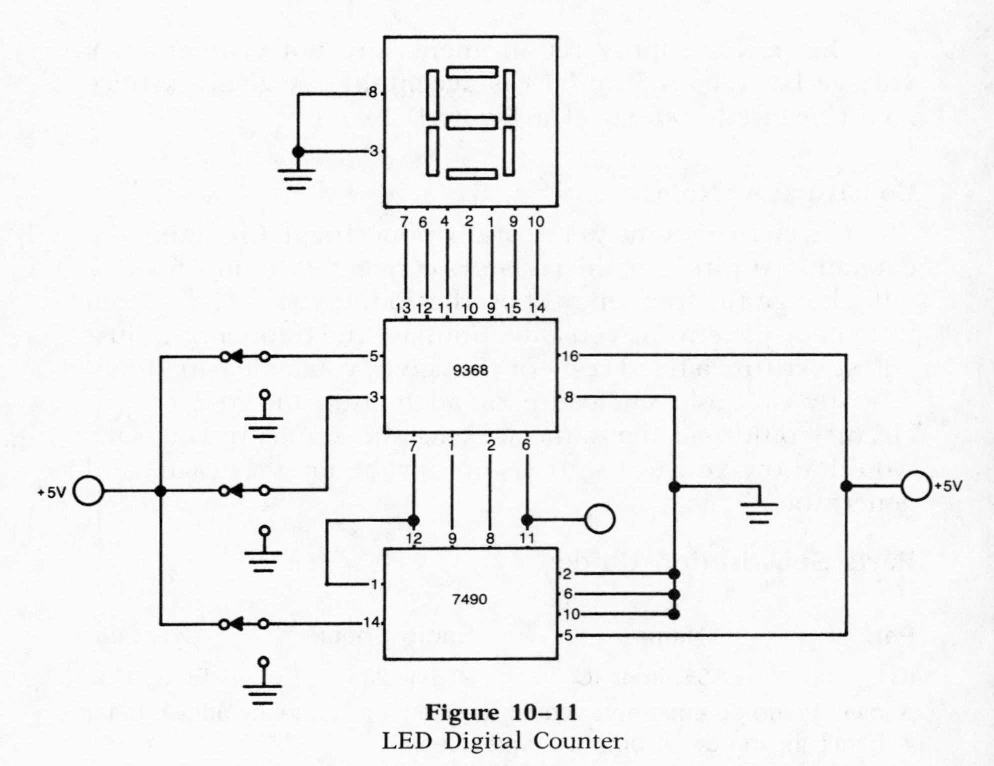

Figure 10–11
LED Digital Counter

When completed, the counter will display the numbers from zero to nine, which are sequenced by pressing the count switch. The counter can be ganged with another to cascade the count function. The maximum digit count is limited only by the current capacity of your power supply.

When interfaced with the timer, an elapsed time counter is easily constructed.

Circuit Description:

The 7490 is a TTL family IC that generates a BCD code. BCD stands for binary coded decimal. The code represents the digits 0 through 9, and is a four-bit code. The counter in the 7490 increments the code from 0000 to 1001 on pins 12, 9, 8, and 11. If you measure these pins, you will see the code change in sequence with the table below.

The 9368 IC is a combination BCD decoder and 7-segment readout driver. The one IC converts the binary code

BCD Truth Table

0000	0	0101	5
0001	1	0110	6
0010	2	0111	7
0011	3	1000	8
0100	4	1001	9

to the proper code to light a common cathode 7-segment LED readout. The circuit operates on a single +5-volt power supply.

Construction Notes:

The use of IC sockets is recommended. The LED will also mount in a standard socket. The project is most easily done using wire wrap, since there are few external components.

To make the circuit multiple digit, allow enough perf board for a three-socket set for each digit. For every digit, duplicate the wiring shown. Connect pin 11 (D) of the ones digit to pin 14 (input) of the tens digit. Continue this process, pin 11 output to pin 14 input for as many digits as you desire. Each circuit will, of course, demand additional current from the power supply.

Pin 5 of the decoder-driver is a leading zero blanker. If you connect a switch to pin 5, and alternate the pin between 0 and 5 volts, you can have switchable zero blanking. For multiple-digit circuits, omit the switch and tie the pin directly on the tens digit and above. Pin 3 on the 9368 is a count-hold pin. Changing the logic on that pin will hold the reading on the display, though the device will still continue to count. Connecting a switch there will allow you to hold a count reading, and gather a current reading by opening and closing the switch again. Experiment with these modes of tying those pins first to +5 then to 0 volts. You will easily be able to see their effects.

You may also want to install a switch on pin 2 that changes the pin status from 0 volts to 5 volts. The pin will reset the counter chip to 0. In multiple-digit circuits, tie all the reset pins together and they can all be reset at once.

Parts Substitution Guide:

Part	Name	Radio Shack	Sylvania
IC 1	7490 BCD counter	276-1808	ECG 7490
IC 2	9368 decoder driver	NA	ECG 8368
LED 1	FND 500 Common cathode	276-1648	ECG 3079

For the record, the ECG number is 8368. Some people might think that is an error in the cross-reference.

The ICs could be replaced with CMOS types—in fact just about any BCD counter or decoder driver would work. Most, however, are not pin-for-pin compatible, and would require rewiring the circuit to match the new pinouts. The advantage of CMOS would be the ability to battery operate the circuit.

Any common cathode LED readout would be usable, as well. Again, pin rewiring will probably be required.

Misc. parts: IC sockets, perf board, mounting hardware, enclosure, etc.

Troubleshooting Tips:

To test the circuit, apply power. The LED should display a random number between zero and nine. If it does not, check power supply wiring and connections to the 9368. An improper count sequence or "garbage" characters indicate incorrect wiring between either the LED and 9368, or the counter and driver ICs. You can use a voltmeter to test the voltages at the output pins. The 7490 can be checked for proper output by comparing it to the truth table on page 289. A 0 is represented by 0 volts, a 1 by +5 volts. These circuits can also be tested using a simple logic probe built from a resistor and LED. Connect the cathode of an LED to ground, and the anode to a 150-ohm resistor. The other end of the resistor can be connected to a test probe. A +5-volt source, or logic 1 will light the LED, while a 0 will not. Using this device, you can go through the circuit testing for count state.

Universal IC Timer (Figure 10–12)

Project Description:

The 555 timer is used in this project as a monostable multivibrator. A monostable circuit will change state for a

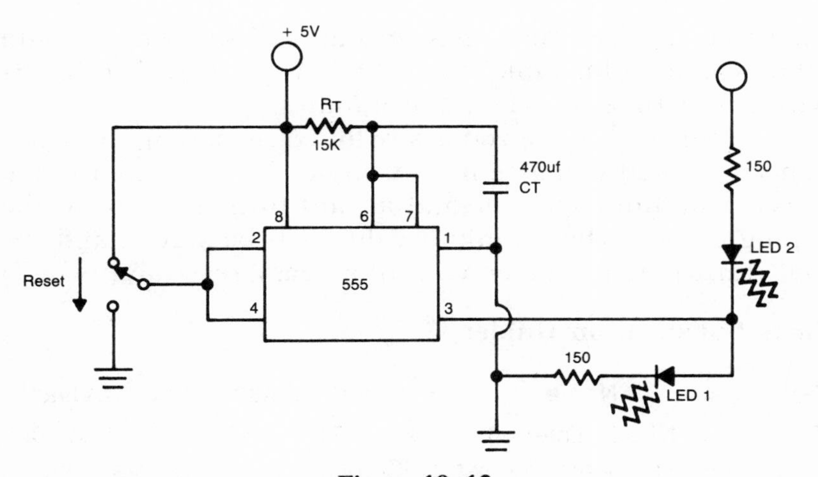

Figure 10–12
Integrated Circuit Timer

certain time period. After the period has elapsed, it returns to its normal state. The circuit shown has a time span of less than 30 seconds. By changing the values of Ct and Rt, you can change the timing function from microseconds to hours. With a little experimentation, you can connect the output of this circuit at pin 3 to the count switch of the counter, and keep track of the number of elapsed time periods.

Circuit Description:

Pins 2 and 4 are connected to a reset pin that restarts the timing cycle. For remote timing purposes, you could design a transistor switch that will trip the timer. The output is connected to two LEDs that will change the state as the timer operates. Our test circuit had different-colored LEDs to class up the operation of the circuit.

Pin 3, the output, can be interfaced to any transistor switch, and will handle up to 200 ma. without needing a switch. The state of pin 3 changes from zero volts to the supply voltage. The circuit will operate with a nine-volt battery if desired.

Construction Notes:

Resistor Rt could be replaced with a multipole switch, which would allow you to install several timing resistors.

Alternatively, a variable resistor could be used, which would allow you an adjustable time delay. A 270 K resistor at Rt will give a time of over three minutes.

Use any technique you desire for construction, the 555 is noncritical and will give little trouble. For accurate timing, use a tantalum capacitor and a multiturn trim pot in the timing circuit. The tantalum value will be stable, and the multiturn potentiometer will allow easy resettability.

Parts Substitution Guide:

Part	Name	Radio Shack	Sylvania
IC 1	NE555 timer IC	276-1723	ECG 955

Misc. parts: Perf board, hardware, IC socket, switch, general purpose LEDs, etc.

Troubleshooting Tips:

Little trouble should be encountered with this circuit, assuming correct wiring. Pin 8 is the positive supply voltage. An electrolytic capacitor will have its positive lead connected to pins 6 and 7. Be sure the LEDs are mounted correctly, the cathodes toward the ground.

If there are no wiring errors, and the IC has supply voltage, it is probably defective.

Computer Programs That Quickly Solve Problems

When you were working with the chapters on mathematics for electronic circuits, you probably made good use of the pocket calculator. Now that you are familiar with some of the formulas required for calculation of electronic circuit parameters, you would like some easier way of calculating them. If you have a home computer that understands BASIC, the following programs will help you when working with those units.

These programs were written in an elementary BASIC. They should run with minor modification on any home computer. We will leave the "window dressing" to you. Format the print statements in any way you like—send them to a printer or use the video screen. Merge them and access them from a menu. Turn them into a large program that will allow you to calculate any desired parameter. In short, modify them to fit your needs and applications.

As a learning aid, a few lines of these programs will be listed as follows:

```
10000 xxxxxxxx
```

When you see a line like this, you can study the program flow, and the material in the preceding chapters. See if you can supply the correct line (or lines). This exercise will help you in remembering the procedures you have learned, and allow you to actually work with the data. As has been said many times, you must understand a process before you can write a program that solves it. This exercise should help you gain better understanding.

At the end of this chapter, there are several lines written to make the program do the proper calculations. Before inserting them, try to write your own algorithm. Remember that there is more than one way to write a program to solve a particular problem, and if your line(s) are different from those listed, don't automatically assume you are wrong, as you have probably solved the algorithm in a slightly different manner. To be sure yours is correct, try your listing with several different sets of data, then using the same data, try the lines listed at the end of the chapter. If the results are the same, your solution is correct. Be sure to try all ranges of data with the program in the event that some data just happen to work properly.

Program List

Solving for Voltage, Current, and Resistance in Series Circuits

Solving for Voltage, Current, and Resistance in Parallel Circuits

Solving for Voltage, Current, and Resistance in Complex Circuits

Solving for Reactance

Solving for Inductance

Solving for Capacitance

Solving for Impedance

Determining Transformer Paramenters

PROGRAM NUMBER 1

Solving for Voltage, Current, and Resistance in Series Circuits

```
10      REM PROGRAM TO SOLVE FOR VOLTAGE, CUR-
          RENT AND RESISTANCE IN SERIES CIRCUITS
20      PRINT
30      PRINT "THIS PROGRAM SOLVES FOR VOLTAGE"
40      PRINT "CURRENT, OR RESISTANCE."
50      PRINT
60      E=0
70      I=0
80      R=0
90      C=0
100     PRINT "ENTER:"
110     PRINT "⟨1⟩ VOLTAGE"
120     PRINT "⟨2⟩ CURRENT"
130     PRINT "⟨3⟩ RESISTANCE"
140     PRINT "⟨4⟩ END PROGRAM"
150     PRINT "YOUR CHOICE"
160     INPUT M
170     IF M=1 THEN GOSUB 1000
180     IF M=2 THEN GOSUB 2000
190     IF M=3 THEN GOSUB 3000
200     IF M=4 THEN GOTO 4000
210     GOTO 50
1000    REM SECTION TO CALCULATE VOLTAGE
1010    PRINT "⟨1⟩ CALCULATE TOTAL VOLTAGE FROM RE-
          SISTANCE AND CURRENT"
1020    PRINT "⟨2⟩ CALCULATE TOTAL VOLTAGE FROM
          VOLTAGE DROPS"
1030    INPUT M
1040    IF M=1 THEN 1140
1050    IF M <> 2 THEN 1000
1060    PRINT "NUMBER OF VOLTAGE DROPS"
1070    INPUT N
1080    PRINT "ENTER VOLTAGE"
1090    INPUT A
1100    XXXXXXXX
1110    C=C+1
1120    IF C<N THEN 1080
1130    GOTO 1250
```

PROGRAM NUMBER 1 continued

```
1140        PRINT "ENTER NUMBER OF RESISTORS"
1150        PRINT "ENTER 1 IF TOTAL RESISTANCE IS KNOWN"
1160        INPUT N
1170        C=C+1
1180        PRINT "VALUE OF RESISTOR"
1190        INPUT A
1200        XXXXXXXX
1210        IF C<N THEN 1170
1220        PRINT "ENTER TOTAL CURRENT IN CIRCUIT"
1230        INPUT I
1240        XXXXXXXX
1250        PRINT "THE TOTAL VOLTAGE IS"
1260        PRINT E
1270        PRINT "VOLTS"
1280        M=0
1290        RETURN
2000        REM SECTION TO SOLVE FOR CURRENT
2010        PRINT "ENTER NUMBER OF VOLTAGE DROPS"
2020        PRINT "ENTER 1 IF TOTAL VOLTAGE IS KNOWN"
2030        INPUT N
2040        C=C+1
2050        PRINT "ENTER VOLTAGE"
2060        INPUT A
2070        E=E+A
2080        IF C<N THEN 2040
2090        PRINT "ENTER NUMBER OF RESISTORS"
2100        PRINT "ENTER 1 IF TOTAL RESISTANCE IS KNOWN"
2110        C=0
2120        INPUT N
2130        C=C+1
2140        PRINT "ENTER RESISTANCE"
2150        INPUT A
2160        R=R+A
2170        IF C<N THEN 2130
2180        XXXXXXXX
2190        PRINT "THE TOTAL CURRENT IS"
2200        PRINT I
2210        PRINT "AMPS"
2220        RETURN
3000        REM SECTION TO SOLVE FOR RESISTANCE
3010        PRINT "ENTER TOTAL CURRENT"
3020        INPUT I
```

PROGRAM NUMBER 1 continued

```
3030    PRINT "ENTER NUMBER OF VOLTAGE DROPS"
3040    PRINT "ENTER 1 IF TOTAL VOLTAGE IS KNOWN"
3050    INPUT N
3060    C=C+1
3070    PRINT "ENTER VOLTAGE"
3080    INPUT A
3090    E=E+A
3100    IF C<N THEN 3060
3110    XXXXXXXX
3120    PRINT "THE TOTAL RESISTANCE IS"
3130    PRINT R
3140    PRINT "OHMS"
3150    RETURN
4000    REM END ROUTINE
4010    PRINT "PROGRAM TERMINATED"
4020    END
```

PROGRAM NUMBER 2

Solving for Voltage, Current, and Resistance in Parallel Circuits

```
10      REM PROGRAM TO SOLVE FOR VOLTAGE, CUR-
          RENT, AND RESISTANCE
20      REM IN PARALLEL CIRCUITS
30      PRINT "THIS PROGAM SOLVES FOR VOLTAGE, CUR-
          RENT"
40      PRINT "AND RESISTANCE IN PARALLEL CIRCUITS"
50      DIM R(10)
60      PRINT
70      E=0
80      I=0
90      R=0
100     C=0
110     PRINT "ENTER:"
120     PRINT "⟨1⟩ VOLTAGE"
130     PRINT "⟨2⟩ CURRENT"
140     PRINT "⟨3⟩ RESISTANCE"
150     PRINT "⟨4⟩ END PROGRAM"
160     PRINT "YOUR CHOICE"
170     INPUT M
```

PROGRAM NUMBER 2 continued

```
180     IF M=1 THEN GOSUB 1000
190     IF M=2 THEN GOSUB 2000
200     IF M=3 THEN GOSUB 3000
210     IF M=4 THEN GOTO 4000
220     GOTO 60
1000    REM SOLVE FOR VOLTAGE
1010    PRINT "ENTER NUMBER OF BRANCH CURRENTS"
1020    PRINT "ENTER 1 IF TOTAL CURRENT AND RESIS-
          TANCE IS KNOWN"
1020    PRINT "ENTER 1 IF TOTAL CURRENT AND RESIS-
          TANCE IS KNOWN"
1030    INPUT N
1040    C=C+1
1050    PRINT "ENTER CURRENT"
1060    INPUT A
1070    I=I+A
1080    IF C<N THEN 1040
1090    C=0
1100    C=C+1
1110    GOSUB 3000
1120    E=I*R
1130    "TOTAL VOLTAGE IN THIS CIRCUIT IS"
1140    PRINT E
1150    PRINT "VOLTS"
1160    RETURN
2000    REM SECTION TO SOLVE FOR CURRENT
2010    PRINT "ENTER NUMBER OF BRANCH CURRENTS"
2020    INPUT N
2030    C=C+1
2040    PRINT "ENTER BRANCH CURRENT"
2050    INPUT A
2060    XXXXXXXX
2070    IF C<N THEN 2030
2080    PRINT "TOTAL CURRENT IN THE CIRCUIT IS"
2090    PRINT I
2100    PRINT "AMPS"
2110    RETURN
3000    REM SECTION TO SOLVE FOR TOTAL RESISTANCE
3010    PRINT "ENTER NUMBER OF RESISTORS IN PARAL-
          LEL"
3020    PRINT "MUST BE 10 OR LESS"
3030    INPUT N
```

PROGRAM NUMBER 2 continued

```
3040        IF N>10 THEN GOTO 3020
3050        XXXXXXXX
3060        XXXXXXXX
3070        XXXXXXXX
3080        XXXXXXXX
3090        XXXXXXXX
3100        XXXXXXXX
3110        XXXXXXXX
3120        XXXXXXXX
3130        XXXXXXXX
3140        PRINT "TOTAL RESISTANCE IS"
3150        PRINT R
3160        PRINT "OHMS"
3170        RETURN
4000        REM END ROUTINE
4010        PRINT "PROGRAM TERMINATED"
4020        END
```

PROGRAM NUMBER 3

Solving for Voltage, Current, and Resistance in Complex Circuits

```
  10        REM THIS PROGRAM WILL CALCULATE VALUES
              FROM COMPLEX CIRCUITS
  20        REM IT WILL THEN GIVE CIRCUIT TOTALS
  30        PRINT "TO USE THIS PROGRAM YOU MUST CAL-
              CULATE"
  40        PRINT "ALL SERIES AND PARALLEL VALUES USING
              THE SERIES AND PARALLEL PROGRAMS"
  50        PRINT "THEN INPUT THOSE NUMBERS INTO THIS
              PROGRAM FOR FINAL CALCULATION"
  60        DIM R(10)
  70        PRINT
  80        E=0
  90        I=0
 100        R=0
 110        C=0
 120        PRINT "ENTER:"
 130        PRINT "<1> VOLTAGE"
 140        PRINT "<2> CURRENT"
```

PROGRAM NUMBER 3 continued

```
150     PRINT "⟨3⟩ RESISTANCE"
160     PRINT "⟨4⟩ END PROGRAM"
170     PRINT "YOUR CHOICE"
180     INPUT M
190     IF M=1 THEN GOSUB 1000
200     IF M=2 THEN GOSUB 2000
210     IF M=3 THEN GOSUB 3000
220     IF M=4 THEN GOTO 4000
230     GOTO 70
1000    REM SECTION TO CALCULATE TOTAL VOLTAGE
1010    PRINT "ENTER TOTAL VOLTAGE FROM SERIES
          ELEMENTS"
1020    INPUT A
1030    E=E+A
1040    PRINT "ENTER TOTAL VOLTAGE FROM PARALLEL
          ELEMENTS"
1050    INPUT A
1060    E=E+A
1070    PRINT "THE TOTAL VOLTAGE IS"
1080    PRINT E
1090    PRINT "VOLTS"
1100    RETURN
2000    REM SECTION TO SOLVE FOR CURRENT
2010    PRINT "TO FIND TOTAL CURRENT, YOU MUST
          KNOW"
2020    PRINT "VOLTAGES AND RESISTANCES OF THE PAR-
          ALLEL AND SERIES BRANCHES"
2030    PRINT "ENTER TOTAL SERIES RESISTANCE"
2040    XXXXXXXX
2050    XXXXXXXX
2060    PRINT "ENTER TOTAL SERIES RESISTANCE"
2070    XXXXXXXX
2080    XXXXXXXX
2090    PRINT "ENTER TOTAL VOLTAGE"
2100    XXXXXXXX
2110    XXXXXXXX
2120    XXXXXXXX
2130    PRINT "THE TOTAL CURRENT IS"
2140    PRINT I
2150    PRINT "AMPS"
2160    RETURN
3000    REM SECTION TO SOLVE FOR TOTAL RESISTANCE
```

PROGRAM NUMBER 3 continued

```
3010    PRINT "ENTER NUMBER OF RESISTOR IN PARAL-
          LEL"
3020    PRINT "MUST BE 10 OR LESS"
3030    INPUT N
3040    IF N>10 THEN GOTO 3020
3050    FOR X=1 TO N
3060    PRINT "ENTER RESISTOR VALUE"
3070    INPUT R(X)
3080    R(X)=1/R(X)
3090    NEXT X
3100    FOR X=1 TO N
3110    R=R+R(X)
3120    NEXT X
3130    R=1/R
3140    PRINT "ENTER NUMBER OF RESISTORS IN SERIES"
3150    PRINT "ENTER 1 IF TOTAL SERIES RESISTANCE IS
          KNOWN"
3160    C=0
3170    INPUT N
3180    C=C+1
3190    PRINT "ENTER RESISTANCE"
3200    INPUT A
3210    R=R+A
3220    IF C<N THEN 3180
3230    PRINT "TOTAL RESISTANCE IS"
3240    PRINT R
3250    PRINT "OHMS"
3260    RETURN
4000    REM END ROUTINE
4010    PRINT "PROGRAM TERMINATED"
4020    END
```

PROGRAM NUMBER 4

Solving for Reactance

```
10      REM THIS PROGRAM SOLVES FOR REACTANCE
20      DIM V(10)
30      PRINT "WHICH REACTANCE DO YOU WISH TO
          SOLVE FOR?"
40      PRINT "⟨1⟩ INDUCTIVE"
```

PROGRAM NUMBER 4 continued

```
  50      PRINT "⟨2⟩ CAPACITIVE"
  60      PRINT "⟨3⟩ TOTAL SERIES"
  70      PRINT "⟨4⟩ TOTAL PARALLEL"
  80      PRINT "⟨5⟩ END PROGRAM"
  90      INPUT M
 100      IF M=1 THEN GOSUB 1000
 110      IF M=2 THEN GOSUB 2000
 120      IF M=3 THEN GOSUB 3000
 130      IF M=4 THEN GOSUB 4000
 140      IF M=5 THEN GOTO 5000
 150      V=0
 160      C=0
 170      GOTO 40
1000      REM SECTION TO SOLVE FOR INDUCTIVE REAC-
            TANCE
1010      PRINT "ENTER FREQUENCY IN HERTZ"
1020      XXXXXXXX
1030      PRINT "ENTER INDUCTANCE IN HENRIES"
1040      XXXXXXXX
1050      XXXXXXXX
1060      PRINT "TOTAL INDUCTIVE REACTANCE IS"
1070      PRINT XL
1080      PRINT "OHMS"
1090      RETURN
2000      REM SECTION TO SOLVE FOR CAPACITIVE REAC-
            TANCE
2010      PRINT "ENTER FREQUENCY IN HERTZ"
2020      XXXXXXXX
2030      PRINT "ENTER CAPACITANCE IN MICROFARADS"
2040      XXXXXXXX
2050      XXXXXXXX
2060      XXXXXXXX
2070      PRINT "TOTAL CAPACITIVE REACTANCE IS NEGA-
            TIVE"
2080      PRINT XC
2090      PRINT "OHMS"
2100      RETURN
3000      REM SOLVE FOR SERIES REACTANCE
3010      PRINT "⟨1⟩ INDUCTIVE OR ⟨2⟩ CAPACITIVE"
3020      INPUT M
3030      IF M=2 THEN 3140
3040      IF M<>1 THEN 3000
```

PROGRAM NUMBER 4 continued

```
3050      PRINT "ENTER NUMBER OF INDUCTORS"
3060      INPUT N
3070      PRINT "ENTER REACTANCE IN OHMS"
3080      GOSUB 6000
3090      XL=V
3100      PRINT "TOTAL INDUCTIVE REACTANCE IS"
3110      PRINT XL
3120      PRINT "OHMS"
3130      RETURN
3140      PRINT "ENTER NUMBER OF CAPACITORS"
3150      INPUT N
3160      PRINT "ENTER REACTANCE IN OHMS"
3170      GOSUB 7000
3180      XC=V
3190      PRINT "TOTAL CAPACITIVE REACTANCE IS NEGA-
            TIVE"
3200      PRINT XC
3210      PRINT "OHMS"
3220      RETURN
4000      REM SUBROUTINE TO FIND PARALLEL REACTANCE
4010      PRINT "(1) INDUCTIVE OR (2) CAPACITIVE"
4020      INPUT M
4030      IF M=2 THEN 4140
4040      IF M <>1 THEN 4000
4050      PRINT "ENTER NUMBER OF INDUCTORS"
4060      INPUT N
4070      PRINT "ENTER REACTANCE IN OHMS"
4080      GOSUB 7000
4090      XL=V
4100      PRINT "TOTAL INDUCTIVE REACTANCE IS"
4110      PRINT XL
4120      PRINT "OHMS"
4130      RETURN
4140      PRINT "ENTER NUMBER OF CAPACITORS"
4150      INPUT N
4160      PRINT "ENTER REACTANCE IN OHMS"
4170      GOSUB 6000
4180      XC=V
4190      PRINT "TOTOAL CAPACITIVE REACTANCE IS NEGA-
            TIVE"
4200      PRINT XC
4210      PRINT "OHMS"
```

PROGRAM NUMBER 4 continued

```
4220    RETURN
5000    REM END ROUTINE
5010    PRINT "PROGRAM TERMINATED"
5020    END
6000    REM SUBROUTINE TO CALCULATE SERIES VALUES
6010    C=C+1
6020    PRINT "ENTER VALUE"
6030    INPUT A
6040    V=V+A
6050    IF C<N THEN 6010
6060    RETURN
7000    REM SUBROUTINE TO CALCULATE PARALLEL
          VALUES
7010    FOR X=1 TO N
7020    PRINT "ENTER VALUE"
7030    INPUT V(X)
7040    V(X)=1/V(X)
7050    NEXT X
7060    FOR X=1 TO N
7070    V=V+V(X)
7080    NEXT X
7090    V=1/V
7100    RETURN
```

PROGRAM NUMBER 5

Solving for Inductance

```
  10    REM THIS PROGRAM SOLVES FOR TOTAL INDUC-
          TANCE
  20    REM IN A SERIES OR PARALLEL CIRCUIT
  30    DIM V(10)
  40    PRINT "WHICH INDUCTANCE DO YOU WISH TO
          SOLVE FOR?"
  50    PRINT "THIS PROGRAM DOES NOT COMPENSATE
          FOR MUTUAL INDUCTANCE"
  60    PRINT "⟨1⟩ SERIES"
  70    PRINT "⟨2⟩ PARALLEL"
  80    PRINT "⟨3⟩ END PROGRAM"
  90    INPUT M
 100    IF M=1 THEN GOSUB 1000
 110    IF M=2 THEN GOSUB 2000
```

PROGRAM NUMBER 5 continued

```
 120     IF M=3 THEN GOTO 3000
 130     V=0
 135     C=0
 140     GOTO 40
1000     REM SOLVE FOR SERIES INDUCTANCE
1010     PRINT "ENTER NUMBER OF INDUCTORS"
1020     INPUT N
1030     PRINT "ENTER INDUCTANCE IN HENRIES"
1040     GOSUB 6000
1050     L=V
1060     PRINT "TOTAL INDUCTANCE IS"
1070     PRINT L
1080     PRINT "HENRIES"
1090     RETURN
2000     REM SOLVE FOR PARALLEL INDUCTANCE
2010     PRINT "ENTER NUMBER OF INDUCTORS"
2020     INPUT N
2030     PRINT "ENTER INDUCTANCE IN HENRIES"
2040     GOSUB 7000
2050     L=V
2060     PRINT "TOTAL INDUCTANCE IS"
2070     PRINT L
2080     PRINT "HENRIES"
2090     RETURN
3000     REM END ROUTINE
3010     PRINT "PROGRAM TERMINATED"
3020     END
6000     REM SUBROUTINE TO CALCULATE SERIES VALUES
6010     XXXXXXXX
6020     XXXXXXXX
6030     XXXXXXXX
6040     XXXXXXXX
6050     IF C<N THEN 6010
6060     RETURN
7000     REM SUBROUTINE TO CALCULATE PARALLEL
           VALUES
7010     FOR X=1 TO N
7020     PRINT "ENTER VALUE"
7030     INPUT V(X)
7040     V(X)=1/V(X)
7050     NEXT X
7060     FOR X=1 TO N
```

PROGRAM NUMBER 5 continued

```
7070    V=V+V(X)
7080    NEXT X
7090    V=1/V
7100    RETURN
```

PROGRAM NUMBER 6

Solving for Capacitance

```
  10    REM THIS PROGRAM SOLVES FOR TOTAL CAPACI-
          TANCE
  20    REM IN A SERIES OR PARALLEL CIRCUIT
  30    DIM V(10)
  40    PRINT "WHICH CAPACITANCE DO YOU WISH TO
          SOLVE FOR?"
  60    PRINT "<1> SERIES"
  70    PRINT "<2> PARALLEL"
  80    PRINT "<3> END PROGRAM"
  90    INPUT M
 100    IF M=1 THEN GOSUB 1000
 110    IF M=2 THEN GOSUB 2000
 120    IF M=3 THEN GOSUB 3000
 130    V=0
 135    C=0
 140    GOTO 40
1000    REM SOLVE FOR SERIES CAPACITANCE
1010    PRINT "ENTER NUMBER OF CAPACITORS"
1020    INPUT N
1030    PRINT "ENTER CAPACITANCE IN MICROFARADS"
1040    GOSUB 7000
1050    C=V
1060    PRINT "TOTAL CAPACITANCE IS"
1070    PRINT C
1080    PRINT "MICROFARADS"
1090    RETURN
2000    REM SOLVE FOR PARALLEL CAPACITANCE
2010    PRINT "ENTER NUMBER OF CAPACITORS"
2020    INPUT N
2030    PRINT "ENTER CAPACITORS IN MICROFARADS"
2040    GOSUB 6000
2050    C=V
```

PROGRAM NUMBER 6 continued

```
2060        PRINT "TOTAL CAPACITANCE IS"
2070        PRINT C
2080        PRINT "MICROFARADS"
2090        RETURN
3000        REM END ROUTINE
3010        PRINT "PROGRAM TERMINATED"
3020        END
6000        REM SUBROUTINE TO CALCULATE SERIES VALUES
6010        C=C+1
6020        PRINT "ENTER VALUE"
6030        INPUT A
6040        V=V+A
6050        IF C< THEN 6010
6060        RETURN
7000        REM SUBROUTINE TO CALCULATE PARALLEL
              VALUES
7010        FOR X=1 TO N
7020        PRINT "ENTER VALUE"
7030        XXXXXXXX
7040        XXXXXXXX
7050        XXXXXXXX
7060        XXXXXXXX
7070        XXXXXXXX
7080        XXXXXXXX
7090        XXXXXXXX
7100        RETURN
```

PROGRAM NUMBER 7

Solving for Impedance

```
  10        REM THIS PROGRAM SOLVES FOR TOTAL SERIES
  20        REM OR PARALLEL IMPEDANCE IN AN AC CIRCUIT
  30        PRINT "TO SOLVE FOR TOTAL IMPEDANCE"
  40        PRINT "YOU MUST KNOW TOTAL RESISTANCE,"
  50        PRINT "AS WELL AS INDUCTIVE AND CAPACITIVE
              REACTANCE"
  60        PRINT "SERIES OR PARALLEL CIRCUIT (S/P)"
  70        INPUT M$
  80        IF M$="S" THEN 110
  90        IF M$<>"P" THEN 60
```

PROGRAM NUMBER 7 continued

```
100   FL=1
110   PRINT "ENTER TOTAL RESISTANCE IN OHMS"
120   INPUT R
130   IF R=0 THEN 150
140   IF FL=1 THEN R=1/R
150   PRINT "ENTER TOTAL INDUCTIVE REACTANCE IN
        OHMS"
160   INPUT XL
170   IF XL=0 THEN 190
180   IF FL=1 THEN XL=1/XL
190   PRINT "ENTER TOTAL CAPACITIVE REACTANCE IN
        OHMS"
200   INPUT XC
210   IF XC=0 THEN 230
220   IF FL=1 THEN XC=1/XC
230   XXXXXXXX
240   IF FL=1 THEN Z=1/Z
250   PRINT "TOTAL CIRCUIT IMPEDANCE IS"
260   PRINT Z
270   PRINT "OHMS"
280   IF XL>XC THEN PRINT "INDUCTIVE"
290   IF XC>XL THEN PRINT "CAPACITIVE"
300   IF XC=XL THEN PRINT "RESONANT"
310   FL=0
320   PRINT "ANOTHER IMPEDANCE (Y/N)?"
330   INPUT M$
340   IF M$="N" THEN 370
350   IF M$="Y" THEN 10
360   GOTO 320
370   PRINT "PROGRAM TERMINATED"
380   END
```

PROGRAM NUMBER 8

Determining Transformer Parameters

```
10    REM THIS PROGRAM SOLVES FOR TRANSFORMER
        RATIOS
20    REM SECONDARY VOLTAGE GIVEN PRIMARY VOLT-
        AGE AND TURNS RATIO
30    REM SECONDARY CURRENT GIVEN PRIMARY CUR-
        RENT AND VOLTAGE RATIO
```

PROGRAM NUMBER 8 continued

```
  40    REM TURNS RATIO GIVEN PRIMARY AND SECOND-
          ARY VOLTAGES
  50    PRINT "(1) SECONDARY VOLTAGE"
  60    PRINT "(2) SECONDARY CURRENT"
  70    PRINT "(3) TURNS RATIO"
  80    PRINT "(4) END PROGRAM"
  90    PRINT "ENTER CHOICE"
 100    INPUT M
 110    IF M=1 THEN GOSUB 1000
 120    IF M=2 THEN GOSUB 2000
 130    IF M=3 THEN GOSUB 3000
 140    IF M=4 THEN GOTO 4000
 150    GOTO 50
1000    REM CALCULATE SECONDARY VOLTAGE
1010    PRINT "ENTER PRIMARY VOLTAGE"
1020    INPUT EP
1030    PRINT "ENTER TURNS RATIO WITH PRIMARY FIRST"
1040    INPUT NP
1050    PRINT "ENTER SECONDARY RATIO"
1060    INPUT NS
1070    XXXXXXXX
1080    PRINT "THE SECONDARY VOLTAGE IS"
1090    PRINT ES
1100    PRINT "VOLTS"
1110    RETURN
2000    REM CALCULATE SECONDARY CURRENT
2010    PRINT "ENTER PRIMARY CURRENT"
2020    INPUT IP
2030    PRINT "ENTER VOLTAGE RATIO WITH PRIMARY
          FIRST"
2040    INPUT EP
2050    PRINT "ENTER SECONDARY VOLTAGE RATIO"
2060    INPUT ES
2070    XXXXXXXX
2080    PRINT "SECONDARY CURRENT IS"
2090    PRINT IS
2100    PRINT "AMPS"
2110    RETURN
3000    REM CALCULATE TURNS RATIO
3010    PRINT "ENTER PRIMARY VOLTAGE"
3020    INPUT EP
3030    PRINT "ENTER SECONDARY VOLTAGE"
```

PROGRAM NUMBER 8 continued

```
3040        INPUT ES
3050        XXXXXXXX
3060        PRINT "TURNS RATIO IS"
3070        PRINT TP
3080        PRINT "TO 1"
3090        RETURN
4000        REM END ROUTINE
4010        PRINT "PROGRAM TERMINATED"
4020        END
```

Solutions to Programs

PROGRAM NUMBER 1

This program tests for knowledge of Ohm's law and Kirchhoff's laws.

Line #	*Solution*	*Remarks*
1100	E=E+A	TOTAL VOLTAGE IS THE SUM OF VOLTAGE DROPS
1200	R=R+A	TOTAL RESISTANCE IS THE SUM OF RESISTOR VALUES
1240	E=I*R	VOLTAGE EQUALS CURRENT TIMES RESISTANCE
2180	I=E/R	OHM'S LAW, AGAIN
3110	R=E/I	THIRD VARIATION OF OHM'S LAW

PROGRAM NUMBER 2

This program tests for knowledge of branch current rules and the parallel resistance formula.

Line #	*Solution*	*Remarks*
2060	I=I+A	VARIABLE A IS USED TO GET VALUES TO BE ADDED TO I
3050	For X=1 TO N	THIS IS A SUBROUTINE TO CALCULATE TOTAL
3060	PRINT "ENTER RESISTOR VALUE"	PARALLEL RESISTANCE
3070	INPUT R(X)	ARRAY R IS USED TO ACCUMULATE TOTAL RESISTANCES
3080	R(X)=1/R(X)	CALCULATE RECIPROCALS OF R AND RETURN TO ARRAY
3090	NEXT X	GET THE NEXT VALUE
3100	FOR X=1 TO N	ADD THE RECIPROCALS TOGETHER
3110	R=R+R(X)	AND STORE THE SUM IN R
3120	NEXT X	GET THE NEXT VALUE
3130	R=1/R	DETERMINE THE RECIPROCAL AND STORE

PROGRAM NUMBER 3

This program tests for knowledge of calculating for complex circuit elements.

Line #	*Solution*	*Remarks*
2040	INPUT A	A IS A TEMPORARY VARIABLE
2050	R=R+A	OBTAIN TOTAL SERIES RESIST-ANCE
2070	INPUT A	
2080	R=R+A	OBTAIN TOTAL RESISTANCE
2100	INPUT A	
2110	E=E+A	OBTAIN TOTAL VOLTAGE
2120	I=E/R	CALCULATE TOTAL CURRENT

PROGRAM NUMBER 4

This program tests for application of the reactance formulas.

Line #	*Solution*	*Remarks*
1020	INPUT F	OBTAIN FREQUENCY
1040	INPUT L	OBTAIN INDUCTANCE
1050	XL=2*3.14159*F*L	CALCULATE REACTANCE
2020	INPUT F	OBTAIN FREQUENCY
2040	INPUT C	OBTAIN CAPACITANCE IN MICROFARADS
2050	C=C*.000001	CONVERT TO FARADS
2060	XC=1/(2*3.14159*F*C)	CALCULATE REACTANCE

PROGRAM NUMBER 5

This program applies the inductance formulas. You are to write the subroutine that calculates a series value for the program to use. Use the variable V to store the total series value.

Line #	*Solution*	*Remarks*
6010	C=C+1	SET COUNTER TO 1
6020	PRINT "ENTER VALUE"	DISPLAY PROMPT
6030	INPUT A	OBTAIN FIRST VALUE

6040	V=V+A	ADD TO VALUE
6050	IF C<N THEN 6010	DO IT AGAIN UNLESS DONE

PROGRAM NUMBER 6

This program asks you to write the subroutine that calculates parallel values. Again, use variable V to store the total.

Line #	*Solution*	*Remarks*
7010	FOR X=1 TO N	COUNT TO NUMBER OF VALUES
7020	PRINT "ENTER VALUE"	PROMPT
7030	INPUT V(X)	STORE VALUE IN ARRAY V
7040	V(X)=1/V(X)	CALCULATE THE INVERSE OF V AND RETURN TO ARRAY
7050	NEXT X	DO AGAIN UNTIL DONE
7060	FOR X=1 TO N	COUNT TO END OF ARRAY
7070	V=V+V(X)	ADD ARRAY AND STORE IN V
7080	NEXT X	DO AGAIN
7090	V=1/V	STORE RECIPROCAL IN VARIABLE BEFORE RETURN

PROGRAM NUMBER 7

This program calculates for total series and parallel impedance. A word of explanation: FL is used as a flag to determine if parallel. If true, the data is converted to reciprocal values before storage.

Line #	*Solution*	*Remarks*
230	Z=SQR((R*R)+((XL−XC)*(XL−XC)))	THIS LINE USES THE SYNTAX "SQR" TO CALCULATE SQUARE ROOT. USE THE SYNTAX THAT FITS YOUR MACHINE. YOU COULD ALSO USE EXPONENTS TO SQUARE.

PROGRAM NUMBER 8

This program calculates transformer ratios.

Line #	*Solution*	*Remarks*
1070	ES=(NS/NP)*EP	CALCULATE SECONDARY VOLTAGE
2070	IS=(EP/ES)*IP	CALCULATE SECONDARY CURRENT
3050	TP=(EP/ES)	CALCULATE PRIMARY TURNS: RATIO ALWAYS USES 1 FOR SECONDARY

Now take a little time and write some of your own programs to assist you in calculating the transistor parameters mentioned in Chapter 8.

Index